Signals from the Subatomic World: How to Build a Proton Precession Magnetometer

Stefan Hollos and Richard Hollos
Exstrom Laboratories LLC
Longmont, Colorado

Abrazol Publishing

an imprint of Exstrom Laboratories LLC
662 Nelson Park Drive, Longmont, CO 80503-7674 U.S.A.

Publisher's Cataloging in Publication Data

Hollos, Stefan
Signals from the Subatomic World: How to Build a Proton Precession Magnetometer / by Stefan Hollos and Richard Hollos
p. cm.
Includes bibliographical references and index.
ISBN 978-1-887187-00-8
Library of Congress Control Number: 2008901596
1. Magnetometers. 2. Nuclear magnetometers. 3. Magnetic instruments. 4. Earth Science instruments. 5. Geomagnetism. 6. Nuclear magnetism.
I. Title. II. Hollos, Stefan.
QC819.H69 2008
681.2 H

This book is dedicated to our parents, Istvan and Anna Hollos, who have helped us in more ways than we can count.

Contents

1 Introduction   1

   1.1 The proton precession magnetometer   2

   1.2 How PPM's work   5

   1.3 Overview of magnetometer 12

2 Polarization Coil and Platform   17

   2.1 Materials not allowed 17

   2.2 Wire for polarizing coil 18

   2.3 Purpose of polarizing coil 19

2.4 Positioning polarizing coil 20

2.5 Inductance of a coil 21

2.6 Magnetic field strength due to coil 22

2.7 Polarizing coil parts list 23

2.8 Specifications of polarizing coil 25

2.9 Construction of polarizing coil 25

2.10 Tiltable platform parts list 26

2.11 Construction of tiltable platform 29

3 Pulse Controller   31

3.1 Purpose of pulse controller 31

3.2 Quenching the current 32

3.3 How a MOSFET works 33

3.4 Advantage of using multiple MOSFETs 34

3.5 Pulse controller circuit description 35

3.6 Microcontroller circuit description 38

3.7 Microcontroller software 39

3.8   Power supply   41

3.9   Assembly and enclosure   42

3.10 Pulse controller parts list   51

4 Sensor Coil   **57**

4.1   Sensor coil requirements   57

4.2   How the coils are wired together   60

4.3   How to connect cable to coils   61

4.4   Specifications of sensor coil   62

4.5   Sensor coils parts list   62

4.6   Construction of sensor coils   63

5 Amplifier   **69**

5.1   Why a differential amplifier is best   70

5.2   Why a low noise amplifier is needed   71

5.3   Description of amplifier circuit   71

5.4   Supplying power to the amplifier   73

5.5   Layout of amplifier circuit   74

5.6 Capacitors at amplifier input 75

5.7 Estimating precession frequency 78

5.8 Amplifier parts list 79

6 Data Acquisition and PC Control   **87**

6.1 What sampling rate to use 88

6.2 What resolution to use 89

6.3 How long to take data 90

6.4 The ADC board used in the Magnum 90

6.5 Pulse controller and data acquisition software . . 92

7 Data Processing and Analysis   **101**

7.1 Signal averaging 101

     7.1.1 Why it improves signal to noise ratio . . . 101

     7.1.2 Limits due to MOSFET slow down 102

7.2 Filtering the data 103

7.3 Spectral analysis using the FFT 108

7.4 Converting bin number to actual frequency . . . 112

7.5   High resolution spectrum for peak location   . . . 113

A Magnum Control Program   **117**

B Microcontroller Program   **129**

Preface

When World War II ended, the physicists at Los Alamos were required to write a report on their work before they could return to their previous lives. This ensured that much of the hard earned knowledge gained at Los Alamos was not lost. In later decades, many benefited from this store of knowledge.

Our Earth's field proton precession magnetometer, called the Magnum, has ended its useful life as a commercial product. And like the administrators at Los Alamos, we, at the risk of sounding pretentious, don't want it's technology to simply be lost in time. Before we move on to other things, we'd rather put it in a book, so other's might find it useful.

The project described in this book is more than just a weekend electronics project for a hobbyist. It would be quite a task for one individual to complete, but still within reach. It's quite suitable for a group project, say for one semester, in a class on instrumentation design, or experimental physics, chemistry or NMR.

To get the most out of this book, it would be good to have had a freshman physics class, and some familiarity with electronics along the lines of Horowitz and Hill's **The Art of Electronics**. Some experience with programming in the C language would also be helpful.

This book also has a website, where much of the contents of the book, and related info can be found:

`http://www.exstrom.com/magnum.html`

We can be reached by email at:
stefan@exstrom.com   richard@exstrom.com

Stefan Hollos and Richard Hollos
Exstrom Laboratories LLC
Longmont, Colorado
Feb 6, 2008

Chapter 1

Introduction

An encounter with the mysterious, that some people never forget, is when they are first given magnets to play with as children. What is this strange force that pulls the magnets together? There is seemingly nothing but empty space between the magnets and yet there is something that either tries to pull them together or push them apart, depending on how they are oriented. This fascinating little phenomena can keep a child busy playing and experimenting for hours. Some children never loose their fascination and go on to become engineers and physicists. That magnetism is indeed a very *deep* phenomena becomes apparent the more you study physics. To fit magnetism into its

proper place in nature requires building up a significant frame-
work of physical theories that includes relativity and quantum
theory.   But it is not necessary to understand magnetism on
such a level in order to get a good intuitive feel for the phenomena and to put it to good use. This book is about using one of
natures smallest magnets, the proton, to measure the strength
of magnetic fields. The device used to do this is called a proton
precession magnetometer and the fact that such a device works
at all is, in our opinion, one of the wonders of nature. Maybe,
we hope, it will inspire you to learn more about the phenomena
of magnetism.

1.1   The proton precession magnetome-
ter

A proton precession magnetometer (PPM) is one of the most
accurate devices that you can build for measuring magnetic
fields. A PPM takes advantage of the fact that protons have
an intrinsic magnetic field, much like an exceedingly small magnet. When the protons are placed in an external magnetic field,
the direction of their field will precess (rotate) about the direction of the external field. The frequency of precession is directly
proportional to the strength of the external field. The propor-
tionality constant, called the gyromagnetic ratio of the proton, is
known to a very high degree of accuracy. When enough protons
are precessing in sync about an external field, they will produce

an oscillating magnetic field, due to the combined effect of each of their precessing fields. The frequency of this oscillating field is ultimately what is measured by the PPM and since it is equal to the precession frequency of the protons it can be used to determine the strength of the external field. This short definitional description of a PPM should become somewhat clearer as the instrument is described in more detail below.

Our goal is to describe, in detail, the construction and operation of a PPM called the Magnum. The Magnum is designed for operation in the Earth's magnetic field or in any magnetic field that is less than about 100,000 nano-Tesla and is uniform over a minimum volume of 1 cubic foot. There are many potential applications for a device like the Magnum. The obvious application is to simply use it to make precision measurements of the Earth's magnetic field. By making periodic measurements, you can create a record of the daily and seasonal variations of the field and monitor magnetic storms. There are a network of magnetic obervatories around the world that continuously monitor the Earth's magnetic field and you can compare your measurements with the that of the closest observatory. The U.S. Geological Survey runs the National Geomagnetism Program which has links to all the magnetic observatories in the United States. There is also an international magnetic observatory network called Intermagnet that has links to magnetic observatories spread throughout the world. If there is no observatory near you then you can contribute your observations to the scientific study of the Earth's magnetic field.

The Magnum can be used to detect magnetic anomalies due to ferromagnetic objects in the environment. The presence of a nearby automobile, building, or any structure with a significant amount of steel, will have an effect on the Earth's field that can be measured with the Magnum. Buried and underwater ferromagnetic objects can also be detected but the spatial resolution will be low. It is possible to modify a Magnum or use more than one of them to improve the spatial resolution but we will leave it to the reader to figure out how to do this.

Calibrating low field magnetometers is another possible application. Magnetometers that use Hall Effect or GMR sensors must be calibrated for them to be effective. The Magnum does not need to be calibrated (the calibration is built into the physical properties of the proton) and so can provide a highly accurate field measurement against which these other magnetometers can be compared. You can for instance callibrate a GMR sensor based magnetometer and then use it for mapping the Earth's field over a large area. This is difficult to do with the Magnum because of its size and weight.

With the Magnum you can measure the nuclear magnetic relaxation time of various proton rich liquids. It thus provides a great introduction to the idea of manipulating the magnetic moments of atomic nuclei. This is the same basic idea that is used in nuclear magnetic resonance (NMR) and magnetic resonance imaging (MRI), except that in these cases the magnetic moments are manipulated in a somewhat more sophisticated manner which we will not discuss here.

The Magnum is also an excellent introduction to scientific instrument design. Basic instrumentation components such as, sensors, signal conditioning, low noise amplification, data acquisition, data analysis, and precision control are all part of the Magnum. In the following chapters the design and construction of the Magnum will be described in detail. In the rest of this chapter we will describe the basic physics behind the operation of the instrument and give a general description of its overall design.

1.2 How PPM's work

To get a feel for what is happening in a proton precession magnetometer lets start by looking at a magnetic compass of the kind that was used for navigation before the GPS came along. If you set the compass on a flat surface, the needle will eventually come to rest, aligned with the Earth's magnetic field. As you probably know, the reason for this is that the compass needle is itself a small magnet and magnets align with the magnetic field that they find themselves in. In this case the north pole of the compass needle is attracted to the magnetic south pole of the Earth [1].

---

[1] The north pole of a magnet is defined as the pole that points toward the geographic north pole of the Earth when the magnet is free to rotate. Since opposite poles attract this means that the magnetic south pole of the Earth is actually near the geographic north pole. We have to say near because the magnetic and geographic poles do not coincide. We should

Earth's magnetic field lines

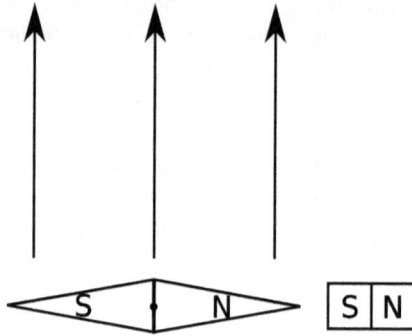

Figure 1.1: Compass near strong magnet

If you now take another magnet, whose field is much stronger than the Earth's, and place it next to the compass, so that its magnetic axis (north-south pole line) is perpendicular to the magnetic axis of the Earth, then the compass needle will realign itself in the direction of the stronger magnet (see figure 1.1). When the stronger magnet is suddenly removed, the compass

---

perhaps also define what we mean by magnetic and geographic poles. The magnetic poles are where the magnetic field lines are oriented vertically. At the north magnetic pole the field lines point vertically out of the Earth and at the south magnetic pole the lines point vertically into the Earth. The geographic poles are where the axis of rotation of the Earth intersects the surface of the Earth. The south magnetic pole is close to the north geographic pole and vice versa.

needle will once again try to align itself with the Earth's field and it will begin to swing back in that direction. When it swings back and reaches alignment it will however not stay there. The needle has acquired some angular momentum during its swing and so it will overshoot the alignment and become misaligned in the opposite direction. The attraction of the Earth's field will however put the brakes on, and make the needle reverse direction back toward alignment.

To cut a long story short, the needle will oscillate back and forth around the direction of the Earth's field, and eventually settle down in alignment with it. An actual compass needle is usually immersed in a fluid that acts to dampen the oscillations so that it will settle down very quickly into alignment with the Earth's field. If the needle were allowed to freely rotate then the oscillations would continue for much longer before settling down. Now if you were to plot the position of the tip of the needle as a function of time you would get a decaying sinusoid as shown in figure 1.2. The interesting thing about this is that it is precisely the same waveform that is produced by a PPM and this more than just a coincidence.

Like a compass needle, the protons that produce the signal in a PPM have an intrinsic magnetic field and for our purposes we can think of them as a bunch of tiny compass needles or magnets. There are however some crucial differences between a proton and a compass needle. To begin with, a proton is not constrained to rotate about a fixed axis like a compass needle. The proton, and hence the direction of its magnetic field, is generally free

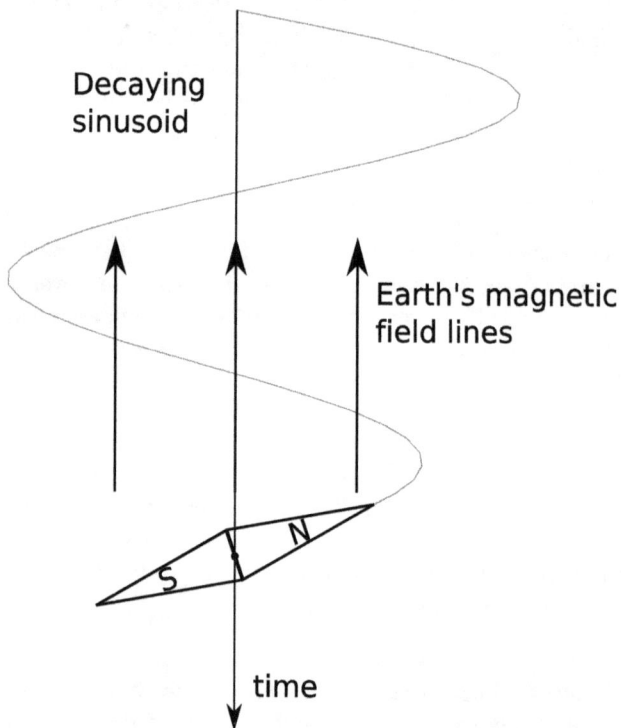

Figure 1.2: Compass needle settling down to alignment with Earth's field.

to rotate in three dimensional space. Another difference is that the proton has an intrinsic angular momentum, in addition to its intrinsic magnetic field. This means that the proton behaves like a small gyroscope or spinning top that also happens to be a magnet.

If you've ever played with a spinning top, you will recall that when the axis of rotation of the top becomes tilted with respect to the vertical (direction of the gravitational force) then the top will start to rotate or precess about the vertical axis. Essentially the same thing happens with a proton when it is placed in a magnetic field. In this case the magnetic field exerts a force on the proton, in the same way that the Earth's field exerts a force on a compass needle, to try to align it with the field. However since the proton is also a spinning top, instead of its field aligning with the external field, it will precess about the direction of the external field just like the spinning top will precess about the direction of the gravitational field.

In a large collection of protons not all of them will be precessing about the magnetic field. The protons interact with each other and with other charged particles such as electrons. The electrons also have intrinsic magnetic fields that can perturb the protons and prevent them from aligning with the field. There will however be a significant number of protons that are precessing about the field and the components of their magnetic fields, that lie in the direction of the external field, will add up to produce some net field. This net field is the origin of what is called nuclear paramagnetism and its strength is proportional to the strength of the external field. [2]

Now look at what happens if we have a bunch of protons in a very strong magnetic field at right angles to a weak field. The strong field will produce some net alignment of the protons in its direction. We say then that the protons have been polarized by the strong field. If the strong field is suddenly set to zero then the only thing that remains is the weak field and the protons will all begin to precess about it in synchronization. This has the effect of rotating the net polarization field produced by the strong field about the direction of the weak field. Recall that this is similar to what happened when the strong magnet was suddenly moved away from the compass needle and the needle then began to rotate back and forth about the weaker Earth field. If an inductive coil is placed around the protons then the rotating polarization field will induce a voltage in the coil which, although extremely weak, can be amplified and measured.

The rotating polarization field will however decay at an exponential rate. This is because the protons become desynchronized so that their fields begin to cancel each other out. The exponential time constant for the decay is called the relaxation time constant and is usually denoted as T_2. The exact value of T_2 will depend on the substance in which the protons are embeded.

---

[2]This is true to a first approximation. There is something similar to feedback that can occur in paramagnetism that will introduce some nonlinearities but it is very weak and can usually be ignored. If these nonlinearities where too large then we would get spontaneous magnetization and the system would be called ferromagnetic and not paramagnetic.

For the Magnum this will usually be some type of proton rich liquid such as water or alcohol. For such liquids T_2 will be on the order of a few seconds and the usable signal will usually be no more than 2 or 3 seconds.

In the above discussion about protons we never specified exactly which protons we were talking about. As you may know, the nuclei of all atoms contain protons and neutrons, both of which have magnetic fields. These magnetic fields are strongly coupled with each other and they combine to produce a net magnetic field for the nucleus. For the purposes of a device such as the Magnum or for any NMR instrument, it is only the net magnetic field of the nucleus that can be can be manipulated by external magnetic fields. The individual protons in a nucleus heavier than hydrogen [3] can not be manipulated in a way that is useful for a PPM.

It is possible in principle to use nuclei heavier than hydrogen in a magnetometer like the Magnum but the signal they produce is too weak to be used in practice. The Magnum is designed to use only the single isolated proton found in the most common isotope of hydrogen. Single protons have the largest magnetic fields and therefore produce the strongest signals. The signal produced by single protons also has the highest frequency and

---

[3]Hydrogen has the lightest nucleus of all the elements. There are three forms, or isotopes, of Hydrogen. The most abundant isotope has only a single proton in the nucleus. The other two isotopes have one or two neutrons in addition to the proton and are relatively rare. When we talk about Hydrogen from here on, it will be assumed that we mean the single proton form.

is therefore the easiest to detect with the inductive sensor used by the Magnum. Another good reason to use single protons is that there are so many compounds that contain hydrogen atoms. Water is the most common but organic compounds such as alcohol and paraffin can also be used. Paraffin is particularly dense in hydrogen and was actually the substance used by Purcell, Pound, and Torrey in their discovery of nuclear magnetic resonance, for which they won the Nobel Prize.

The above discussion should give you a general idea of what a PPM is and the basic principles behind how it operates. If you are interested in learning more about magnetism and nuclear magnetic resonance see the Bibliography at the end of this book. What follows is a brief overview of the operation and design of the Magnum. The individual components of the instrument will be described in detail in the following chapters.

1.3 Overview of magnetometer

First a general description of how to operate the Magnum. You begin by selecting some hydrogen rich substance such as water, alcohol, or paraffin, which we will refer to from here on as the sample. You place about two ounces (59 ml) of the sample inside a cylindrical plastic bottle. The bottle is placed inside the sample coil which is then placed inside the polarization coil. The sample is polarized for a period of time with the field generated by the polarization coil. When the polarization coil is switched off, the signal generated by the precessing protons in the sample is recorded by the data acquisition system. The digitized signal is then processed using software to produce a spectrum which gives the precession frequency of the protons.

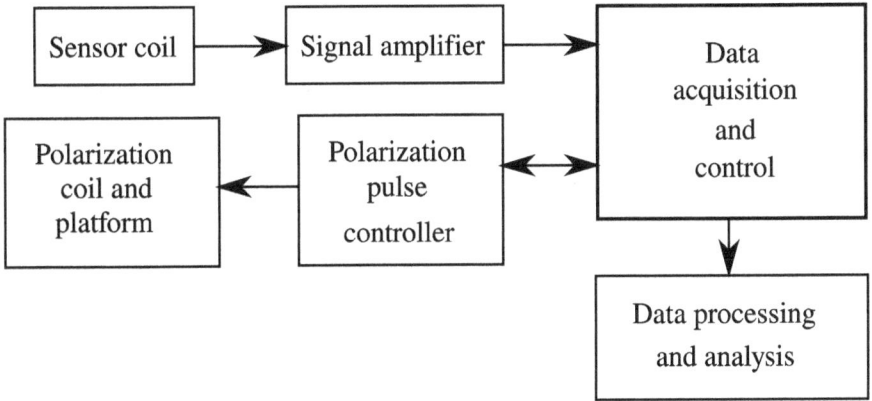

Figure 1.3: Magnum system diagram.

Now we give a description of the overall design of the instrument. A general system diagram is shown in figure 1.3. The Magnum can roughly be divided into three parts: the polarization system, the sensor system, and the control and data acquisition system. We start with the polarization system.

The purpose of the polarization system is to generate a large magnetic field (relative to the Earth) that is in a direction perpendicular to the Earth's field. The polarization field is generated by a large coil which is mounted on a tiltable platform. The tiltable platform is needed to allow the direction of the polarization field to be aligned perpendicular to the Earth's field. The polarization field must be capable of being turned on for a specific period of time and then turned off very quickly. The polarization pulse controller is designed for this purpose. The controller can be programmed by the operator via a serial link to a personal computer. Once programmed, the controller waits for a signal from the PC to start the polarization process. This process consists of turning on the current to the polarization coil for the specified period of time, turning off the current quickly, and then signaling the data acquisition system to begin taking data. The polarization coil, tiltable platform, and pulse controller will be described in detail in subsequent chapters.

The sensor system consists of two parts, the sample coil and the amplifier. The sample coil holds the bottle containing the sample. Its purpose is to inductively pick up the very small oscillating magnetic field produced by the polarized protons precessing about the Earth's magnetic field. The sample coil actually consists of two identical coils placed side by side. One coil is used to house the sample and the other is used to cancel out environmental magnetic fields. Either one of the coils can be used to house the sample but you cannot put a sample in both coils since the two signals would cancel out. The signal produced by the precessing protons is extremely weak. To boost the signal, the sample coil must be tuned using capacitors to form a resonant circuit at the proton precession frequency. The amplifier has a bank of capacitors that can be used to tune the sample coil. The amplifier will also amplify the signal from the tuned sample coil by a factor of about 3.8 million. An amplifier with such a high gain is a very sensitive device and it must be constructed very carefully to prevent it from turning into an oscillator. Details of the construction of the amplifier and the sample coil can be found in the following chapters.

The data acquisition and control system consists of a personal computer with a pci data acquisition card and some software. The software allows the operator to program the pulse controller and then start the polarization and data acquisition process. Multiple cycles of polarization and data acquisition can be controlled by the software. The data from each cycle is averaged together to get a higher signal to noise ratio. The software can program the pulse controller to turn on the polarization field for anywhere from a fraction of a second to several seconds. After it sends a signal to the pulse controller to start the polarization, it initializes the data acquisition system to wait for a signal back from the controller indicating that the polarization has finished and data acquisition can start. Channel selection, setup and sampling rate of the data acquisition system is also controlled by the software. After all the data has been collected, additional software is used to filter the data and calculate the spectrum of the signal. The spectrum then gives the proton precession frequency from which the magnetic field can be calculated. All of this will be described in more detail in the following chapters.

Chapter 2

Polarization Coil and Platform

2.1 Materials not allowed

The polarization coil and tiltable platform upon which it sits must not contain any ferromagnetic material, since this would alter the natural uniformity of the Earth's field, resulting in varying precession frequencies of the protons within the sample, and weakening the signal induced in the sample coil. For the same reason, care must be taken that there are no ferromagnetic materials under or near the polarization coil and platform, and the polarization coil/sensor coil should be placed outside of buildings, when in operation, since most buildings contain large amounts of ferromagnetic materials in their infrastructure (steel I-beams, airducts, nails, etc).

One of our customers once asked whether placing the entire coil assembly inside of a large nonferromagnetic metal box would eliminate magnetic field noise. This is definitely NOT a good idea. Doing so would induce eddy currents in the box when the polarization coil current is turned off, corrupting the magnetic field in the region of the sample, and dashing all hope of getting a signal in the sample coil. For this reason, you want to keep away from the polarization coil any large metallic objects, whether ferromagnetic or not.

2.2 Wire for polarizing coil

A current of approximately 10 amps is sent through the polarizing coil to polarize the sample. The wire size used for the polarizing coil must be large enough to handle this current. We use a 14 gauge, 600 volt THHN insulated solid copper wire for the polarizing coil, and a 16 gauge, 300 volt flexible stranded lamp cord for the 50 ft (15 meters) long connection from the polarizing coil to the pulse controller. The slightly thinner lamp cord is OK since it is not wound like the coil wire and doesn't have much more resistance, but if you can find a 14 gauge lamp cord, you can use that. Note that this would slightly increase the current through the polarizing coil.

The pulse controller turns off the approximately 10 amps of current going through the polarizing coil in about 300 milliseconds. This quick turnoff induces a large voltage of about 270 volts. Therefore the voltage rating of the wire used in constructing the polarizing coil is important. Note that we use a 600 volt wire for the polarizing coil itself, and a 300 volt wire for connection to the coil. Of course you can use a higher voltage wire than this. Keep in mind that the voltage rating of a wire is solely dependent on the wire's insulation, not the wire itself.

2.3 Purpose of polarizing coil

The purpose of the polarizing coil is to produce a magnetic field much greater than that of the Earth's, so that the majority of the population of protons in the sample are precessing about the field of the coil and not the Earth's field. For this reason, the field produced by the polarizing coil need not be especially uniform, as long as it is large compared to the Earth's field. In fact this property inspired the name *Magnum* since it is a brute force method of manipulating the spins of nuclei.

2.4 Positioning polarizing coil

The symmetry axis (the axis around which the coil is wound) of the polarization coil should be oriented so that it is perpendicular to the Earth's magnetic field. In the northern hemisphere, the magnetic field enters the ground at an angle called the magnetic inclination The polarization coil must therefore be tilted from the horizontal by an amount equal to the magnetic inclination angle. So to properly align the coil, orient the platform on which the coil is mounted (using a standard magnetic compass), such that the axis about which the coil tilts, is perpendicular to the magnetic north direction, and tilt the coil by an amount equal to the magnetic inclination. The coil is held in the tilted position by two brass strips that slide onto rods on either side of the platform. The holes in the brass strips allow the angle to be fixed at approximately five degree increments from zero to ninety degrees.

The following websites provide geomagnetic data to allow magnetic inclination angle to be calculated at most locations on the Earth:

- USGS Observatories,
  http://geomag.usgs.gov/observatories/

- INTERMAGNET Participating Observatories,
  http://www.intermagnet.org/Imomap_e.html

- DGRF/IGRF Geomagnetic Field Model,

http://modelweb.gsfc.nasa.gov/models/igrf.html

As an example, the magnetic inclination angle in Longmont, Colorado is approximately 65 degrees. The polarization coil must therefore be aligned as shown in figure 2.1.

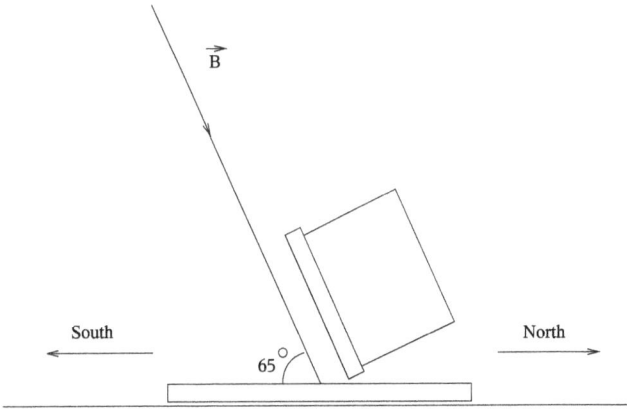

Figure 2.1: Magnetic inclination angle and coil alignment in Longmont, Colorado.

2.5 Inductance of a coil

Knowing the approximate inductance of a coil is necessary, for example, to calculate the voltage induced when the current is turned off, and later, for calculating the resonant frequency of a sample coil, given a capacitance. The inductance of a coil can be approximated with the following formula [20, p42-44].

$$L = \frac{4\pi^2 r^2 N^2 k_n}{l} 10^{-9} \, H$$

where

$$k_n = \frac{1}{1 + 0.9\frac{r}{l} + 0.32\frac{t}{r} + 0.84\frac{t}{l}}$$

and r = mean radius [cm], t = thickness of windings [cm], l = length of coil [cm], N = total number of turns.

2.6 Magnetic field strength due to coil

It is important to know approximately the strength of the magnetic field inside the polarization coil, so that one knows that the field is sufficiently large to get a proton precession signal. The magnetic field, in units of Tesla, due to a multilayered finite length solenoid is

$$B(x) = (\frac{\mu n^2 I}{2}) \left[(\frac{h}{2} - x) \ln(a(x)) + (\frac{h}{2} + x) \ln(b(x)) \right]$$

where x is the distance from the geometric center of the coil along its axis, $\mu = 4\pi * 10^{-7}$ Tesla*meters/Amperes, n is turns/meter of the wire, I is the current through the solenoid in Amperes,

h is the length of the coil in meters, ln is the natural logarithm (log base e), and

$$a(x) = \frac{r_2 + \sqrt{r_2^2 + (h/2 - x)^2}}{r_1 + \sqrt{r_1^2 + (h/2 - x)^2}}$$

$$b(x) = \frac{r_2 + \sqrt{r_2^2 + (h/2 + x)^2}}{r_1 + \sqrt{r_1^2 + (h/2 + x)^2}}$$

where r_1 and r_2 are the inner radius and outer radius, respectively, of the coil in meters.

What is the minimum field strength needed from the polarizing coil in order to get a good signal out of the sample coil? The short answer is, the larger the better, since a stronger field produces a stronger signal, but of course there are practical limits. For the specifications given in this book, you should have about 75 Gauss in the geometric center of the polarizing coil, i.e. $B(0) \approx 0.0075$ in the formula above.

2.7 Polarizing coil parts list

- Acrylic tube, 4.5 inch (11.43 cm) outer diameter, 4 inch (10.16 cm) long, 1/8 inch (0.3175 cm) thick.

- 2 Acrylic squares (end pieces), 6 inch (15.24 cm) on a side, 5/32 inch (0.397 cm) thick.

- 14 AWG solid copper wire, 289 feet (88.1 meter).

- 14 AWG copper speaker wire, 50 feet (15.2 meter).

- Spade Terminals for 14 AWG wire, Qty 2, Mouser Part#538-19144-0015.

Figure 2.2: Wound polarizing coil form.

2.8 Specifications of polarizing coil

- 14 AWG solid copper wire, wire diameter(w/o insul) = 1.63 mm

- coil inner diameter = 11.4 cm

- coil length = 10.0 cm

- number of wire layers = 6

- turns/layer = 38 =¿ total number of turns = 228

- thickness of windings = 1.43 cm

- Approximate inductance = 4.3 mH

- Approximate resistance = 1.1 Ohm

- Approximate self capacitance = 570 pF

2.9 Construction of polarizing coil

The polarizing coil form is composed of three parts, all made of acrylic (also called plexiglass): the tube, and the two end pieces. The two end pieces, which are squares 6 inches (15.24 cm) on a side, need to have circular holes with a diameter of 4 1/4 inches (10.795 cm) drilled into their centers. Keep one of the discs left over from drilling the holes. It will be used later for attaching to the sample coil. Four small holes need to be drilled near each of the corners of one end piece so that the polarizing coil can be attached to the tiltable platform. Acrylic solvent can then be used to glue the end pieces onto the tube. The acrylic and solvent can be purchased at a plastics supplies store.

2.10 Tiltable platform parts list

- Top, white acrylic, 0.5 x 6 x 6 inches (1.27 x 15.24 x 15.24 cm).

- Bottom, white acrylic, 0.5 x 6 x 15 inches (1.27 x 15.24 x 38.1 cm).

- 2 Acrylic hinges.

- Brass strips, 1/4 x 1/16 x 12 inches (0.635 x 0.159 x 30.5 cm).

- Brass rod, diameter 3/32 inches (0.238 cm).

- 4 Nylon machine screws, 8-32 x 1 inch (2.54 cm).

- 4 Nylon nuts.

- 4 Nylon washers.

- 2 rubber bumpers (attach to the bottom of the top piece of acrylic to separate the two pieces).

Figure 2.3: Tiltable platform with polarizing coil mounted on it.

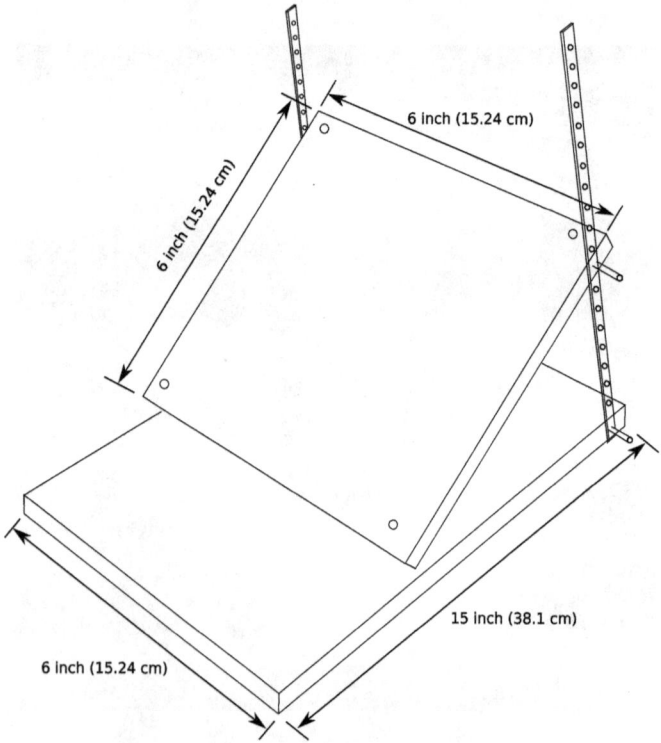

6 inch (15.24 cm)

6 inch (15.24 cm)

6 inch (15.24 cm)

15 inch (38.1 cm)

6 inch (15.24 cm)

Thickness of top and bottom acrylic pieces is 0.5 inch (1.27 cm)

Figure 2.4: Tiltable platform drawing.

2.11   Construction of tiltable platform

The top piece of the tiltable platform needs to have four holes drilled into each of its four corners, so that the polarizing coil can be mounted on it, using the nylon machine screws.

Two acrylic hinges are used to physically attach the top and bottom pieces together. The arylic hinges are glued to the two pieces using acrylic solvent.

Holes are drilled into the sides of the platform, opposite the hinges, so that brass rods, can be inserted. The holes should be just wide enough for the rods to snuggly slide in, and should be about 1/2 inch (1.27 cm) deep. Cut 4 rods about 1 inch (2.54 cm) long, and slide them into the holes with a dab of glue so that they won't ever come out, and so that they protrude out about 1/2 inch. These brass rods are for attaching the brass strips, so that the platform angle can be adjusted. The two brass strips need to have holes drilled into them so that the platform angle can be adjusted in 5 degree increments.

Chapter 3

Pulse Controller

3.1 Purpose of pulse controller

The pulse controller is used to turn the current in the polarization coil on and off. The time it takes to turn on the current is not critical but the time it takes to turn it off is. To get a good precession signal, the polarization field has to be turned off as fast as possible. If the polarization field decays slowly then the proton magnetic moments will simply follow the field and you will end up with no precession signal.

Quickly turning off a large amount of current in a large coil is a tricky business. There is a significant amount of energy stored in the magnetic field of the coil and this energy has to go somewhere. The problem is entirely analogous to trying to stop a large truck moving at high speed in as short a distance as possible without damaging it.

3.2   Quenching the current

You can probably think of many ways of trying to tackle this problem. The simplest solution that immediately comes to mind is to use a mechanical relay. There are several problems with this. First of all, a mechanical device like a relay has a relatively short lifetime compared to solid state switching devices like transistors. In a PPM like the Magnum, the current will be switched on and off so many times that a mechanical relay will fail rather quickly.

Another problem with relays is that they do not turn on and off cleanly and predictably. A relay is basically just a switch like the light switch is your house and the contacts, like any switch will tend to stick when you open them and bounce when you close them. A final problem is the possibility of electric arcing across the contacts when you open them due to the large voltage induced in the coil when the current is turned off.

The problem of quickly turning off the current in a coil has been around for a long time and there was really no good reliable solution until the invention of semiconductor power devices. Some of the early circuits using these devices where composed of a complex and sometimes ingenious combination of zener diodes, resistors, large capacitors, and silicon controlled rectifiers (SCR's). Semiconductor power devices have evolved a great deal since then and a much simpler solution is now possible using a device called a power MOSFET.

3.3 How a MOSFET works

The type of MOSFET used in the magnum is called a p-channel enhancement type. It is basically a solid state switch. The device has three terminals called the source, drain and gate. The gate voltage controls the current flow through the device from the source to the drain [1]. In a p-channel MOSFET the gate to source voltage must be negative to allow source to drain current to flow. The lower the gate voltage is with respect to the source voltage, the more current will flow through the device, up to a certain point.

Like all real devices, a MOSFET has limits. The MOSFETs used in the pulse controller are IRF6215's made by International Rectifier. They have a maximum allowable gate to source volt-

---

[1]Descriptions of current flow are in terms of conventional current. This means that positive current flows from a high potential to a lower potential. This is opposite to the flow of electrons which always move from a low potential to a high potential.

age of -20 volts. The amount of current that can flow through the device and the amount of power it can dissipate is also limited. Another important factor to take into account is the drain to source breakdown voltage. This is the voltage at which current will be conducted through the device even though it is in the off state. This is similar to what happens in a mechanical switch, when the voltage between the contacts gets too high you will get an electrical arc and current flow even though the switch is open.

3.4   Advantage of using multiple MOS-FETs

Fortunately, some of the limitations of a single MOSFET can be overcome by using several of them together. You can increase the effective drain to source breakdown voltage by putting two of them in series. This will double the break down voltage. A problem with this however, is that a MOSFET with current flowing through it will always have a voltage drop across it and therefore some finite resistance. This resistance is called the on-resistance and if you put two MOSFETs in series you effectively double the on-resistance.

From elementary circuit theory you know that putting two resistors of equal value in parallel will give you an equivalent resistance of half the value. The solution to the problem of the doubling of the on-resistance is then to use a parallel combination of two MOSFETs in series for a total of four MOSFETs. This will give you a doubling of the breakdown voltage without increasing the on-resistance.

3.5 Pulse controller circuit description

The schematic for the Magnum pulse controller circuit is shown in figure 3.1. Note that two parallel combinations of four MOSFETs are placed in series. A parallel combinaton of four MOSFETs gives an effective on-resistance of one fourth the on-resistance of a single MOSFET. Putting two of these combinations in series will then result in an overall effective resistance of one half the on-resistance of a single MOSFET. The two combinations in series will also double the effective breakdown voltage. The IRF6215 has an on-resistance of about 0.3 ohms [2] and a minimum breakdown voltage of about 150 volts. With the eight IRF6215s arranged as in the Magnum circuit the effective on-resistance is then 0.15 ohms and the breakdown voltage is 300

---

[2]The on-resistance is a function of the gate to source voltage, the current through the device, and the temperature. The value of 0.3 ohms is typical of the way the IRF6215 is used in the Magnum. It is important to keep in mind however, that the on-resistance will increase as the temperature of the device increases. When the on-resistance increases the current delivered to the polarization coil will decrease. If the temperature of the device increases substantialy it could lead to a significant decrease in the polarization of the protons and loss of precession signal strength. Keep this in mind when you are using a sequence of long polarization pulses.

volts.

A positive twelve volt power supply is connected directly to the source terminals of the first parallel combination of IRF6215's and the polarization coil is connected directly to the drain terminals of the second parallel combination of IRF6215's. When the IRF6215's are turned on we then effectively have a 12 volt supply in series with the 0.15 ohm resistance of the IRF6215s, and the 1.1 ohm resistance and 4.3 mH inductance of the polarization coil. This will deliver approximately $12/1.25 = 9.6$ amps of current [3] to the polarization coil.

The level of the gate voltage on all eight of the IRF6215's is controlled by an Atmel AT90S2313 8-bit microcontroller. To avoid potential damage to the AT90S2313, due to the large induced voltages during switching, the gate voltage is actually controlled through a 4N35 optocoupler. When the output of pin PB0 on the AT90S2313 is high the transistor on the output side of the 4N35 is turned off. This puts the gate voltage of the IRF6215's at approximately 12 volts so that the gate to source voltage is about zero and the devices are turned off. When pin PB0 goes low the output transitor of the 4N35 is turned on, which causes the gate voltage of the IRF6215's to drop to about 1 volt. The gate to source voltage is then approximately -11 volts and the IRF6215's are turned on, allowing the polarization current to flow through the coil.

---

[3]Due to tolerances of circuit components all figures that we give are approximate. To find the actual amount of polarization coil current or any other figure that we give in this book, you will have to measure it.

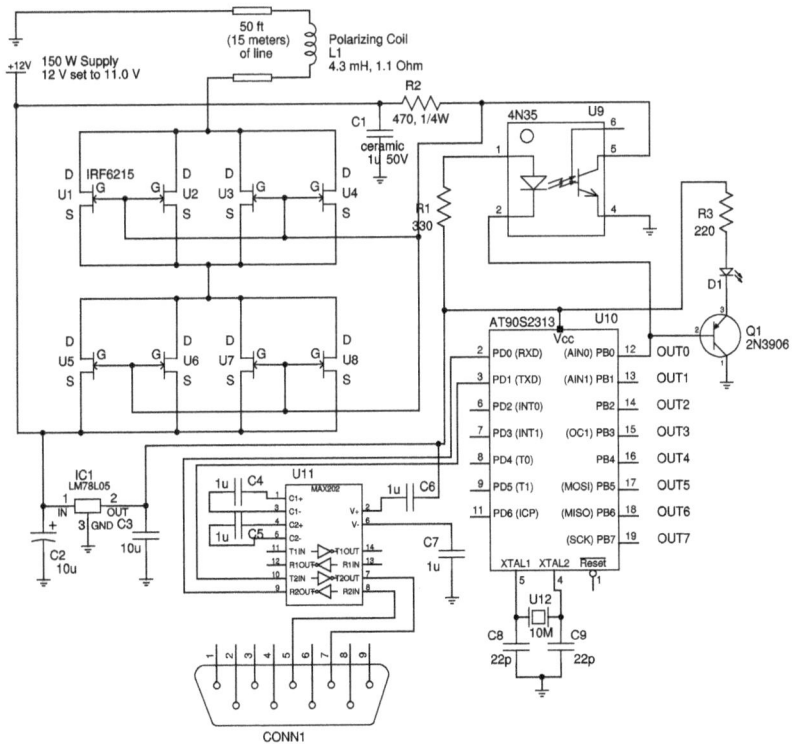

Figure 3.1: Magnum pulse controller and microcontroller circuits.

3.6 Microcontroller circuit description

The voltage supply for the AT90S2313 microcontroller is provided by an LM78L05 voltage regulator that takes the same 12 volt supply used to power the polarization coil, and reduces it to the 5 volts needed by the AT90S2313. The clock speed of the AT90S2313 is set to 10 MHz by the crystal and capacitors connected to pins XTAL1 and XTAL2. Pin PB0 of the AT90S2313, which controls the gate voltage of the MOSFETs is also connected to a 2N3906 transistor that drives an indicator LED which turns on when current is being applied to the polarization coil. All 8 pins of port B on the AT90S2313 are connected to a terminal block on the front of the pulse controller housing. Each of these pins can be individually programmed to turn off and on for specified periods of time as described in the section below on the microcontroller software.

Pins PD0 and PD1 of the AT90S2313 are connected to a MAX202 which is an RS232 driver IC. This IC provides an interface to an RS232 serial port. The serial port is used to allow a PC to communicate with the AT90S2313. The PC can send a pulse program to the AT90S2313 which tells it how long to turn on the polarization current, and then how long to wait before signaling the data acquisition system to start taking data, after the polarization current has been turned off. After the AT90S2313 has been programmed it waits for a signal from the PC to run its program.

3.7 Microcontroller software

The microcontroller software is written in the assembly language of the AT90S2313. The software listing is given in Appendix B. What follows is a general description of how the microcontroller is programmed to behave.

The AT90S2313 continuously listens on the RS232 port for commands from the PC. Its software is completely interrupt driven so that it can respond to commands even during the execution of the pulse program. The five commands that it recognizes are

1. Set Program: the PC transmits this command when it wants to send a new pulse program. After receiving this command the AT90S2313 waits to recieve a new program from the PC which it then stores in memory for later execution.

2. Get Program: the PC transmits this command when it wants the AT90S2313 to send it the pulse program that it has stored in memory. This is useful for verifying that the stored program is correct and has not been corrupted somehow durring transmission to the AT90S2313.

3. Run Program: the PC transmits this command when it wants the AT90S2313 to start executing the pulse program that it has stored in memory. The AT90S2313 will execute the program and then once again go into an idle state where it listens for the next command.

4. Loop Program: this is like the Run Program command except that it tells the AT90S2313 to run the pulse program in a continuous loop. After executing the program once it immediately begins executing it again and it continues doing this until it receives another command.

5. Stop Program: this command tells the AT90S2313 to immediately stop what it is doing and go into an idle state where it just listens for another command from the PC.

The program for the AT90S2313 can be represented as a series of states that the device cycles through. A state is specified by the value of the 8 pins on port B of the device. The value of these 8 pins can be represented by a 8 bit binary number. A low value on one of the pins is taken to be a logic 1, or the on-state, and a high value is taken to be a logic 0, or the off-state. If for example pins PB0 and PB2 are low and all the other pins are high, this can be represented by the binary number 00000101 = 5. To specify a state with all pins set low (logic 1, on-state), use the value 255 = 11111111.

The length of time that a state remains active is specified by two additional bytes. If t is the length of time in units of micro-seconds (1000000 micro-seconds = 1 second), that the state should remain active, then the value of these two bytes is given by the formula:

$$65536 - \text{int}(5 * t / 512)$$

where int() means take the integer part of the argument. The value of t can range from 102.4 to 6710886 micro-seconds which is 0.0001024 to 6.710886 seconds.

A state is then completely specified by 3 bytes. The first byte gives the value of the output pins and the next two bytes give the time that the state remains active or on. A program can have a maximum of 10 states. The AT90S2313 remains in a state for the specified period of time and then transitions to the next state. If the device is in loop mode when the time runs out on the last state, it will loop back to the fist state and continue, otherwise it will remain in the last state. It is important to keep in mind that the device remains in the last state when the program finishes unless it is in loop mode.

The process of actually programing the AT90S2313 is handled by software running on a PC. The software allows the states and times to be specified very easily in a parameter file and it calculates the necessary byte values so that you don't actually need to use the above equation. This software will be discussed in a following chapter.

3.8 Power supply

The power supply used for supplying current to the polarizing coil and to the microcontroller circuit (through an LM78L05) is a 12 volt, 150 watt, Meanwell power supply (Manufacturer#S-150-12). Other power supplies might be usable, but the two important properties that must be considered are the power, and the ability to withstand transients at the output. Another good thing about the Meanwell power supply is that it fits inside the enclosure specified in the parts list.

3.9 Assembly and enclosure

Figure 3.2 shows the physical layout of the MOSFETs. All 8 MOSFETs are mounted on a single aluminum bar (5.6 inch (142mm) long x 1.0 inch (25.4 mm) wide x 1/8 inch (3.175 mm) thick), with thermal pads (see parts list) between the MOSFETs and the bar. The aluminum bar is, in turn, attached to the enclosure of the pulse controller. This provides a good heatsink, and ties all the MOSFETs together thermally. The fact that the MOSFETs have a common thermal connection is important because the speed at which MOSFETs switch is temperature dependent. Their switching speed slows as they get hotter.

Because of the large currents, the MOSFET switching circuit is not implemented on a printed circuit board, but is constructed on a perfboard with wires connecting components. The **thick** wires shown in Fig. 3.2 carry the approximately 10 amps of current, so there we use a heavier 12 AWG stranded wire. Of course the wires from the power supply to the MOSFET circuit also need to be a heavier 12 AWG. The thinner wires between the gates of the MOSFETs and the optocoupler (4N35) can be regular breadboard gauge wire.

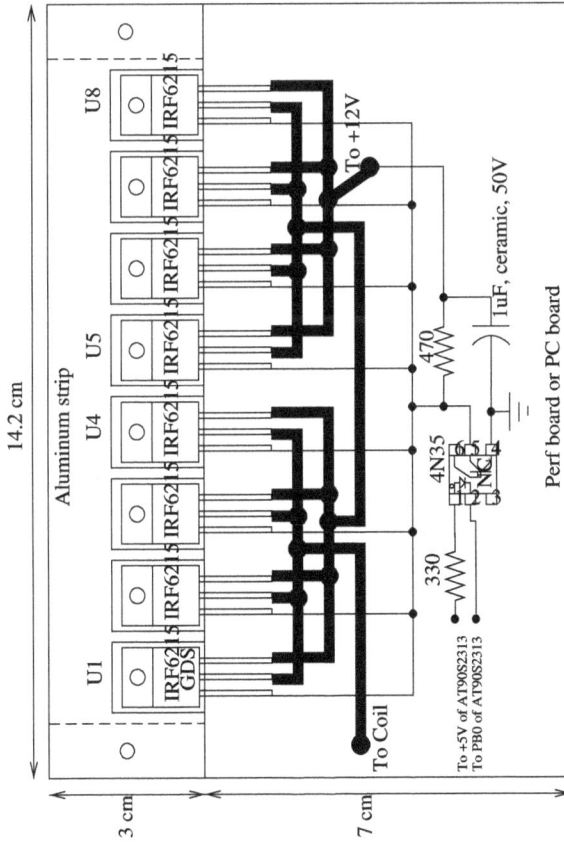

Figure 3.2: Physical layout of the MOSFETs.

The thermal pads between the MOSFETs and the aluminum bar
not only provide a thermal connection for the MOSFETs but
also provide electrical isolation between the MOSFETs and the
enclosure. This is necessary because the metal tab, normally
used for mounting the MOSFETs, is internally connected to
the drain. For this reason, we also use nylon machine screws to
mount the MOSFETs.

A number of holes need to be drilled/cut into the enclosure.

- Two holes for the binding post on front.

- Two holes for the LED's on front.

- Holes for the terminal block on front.

- Power entry module on back.

- RS232 connector on back.

- MOSFETs and aluminum bar on back.

- Power supply on bottom.

Note that the microcontroller board has a RS232 connector
mounted on it, and that connector is mounted onto the enclosure from the inside. That's all the microcontroller board needs
for mounting, although you should put a rubber bumper under-
neath the board to prevent it from accidentally shorting out on
the enclosure.

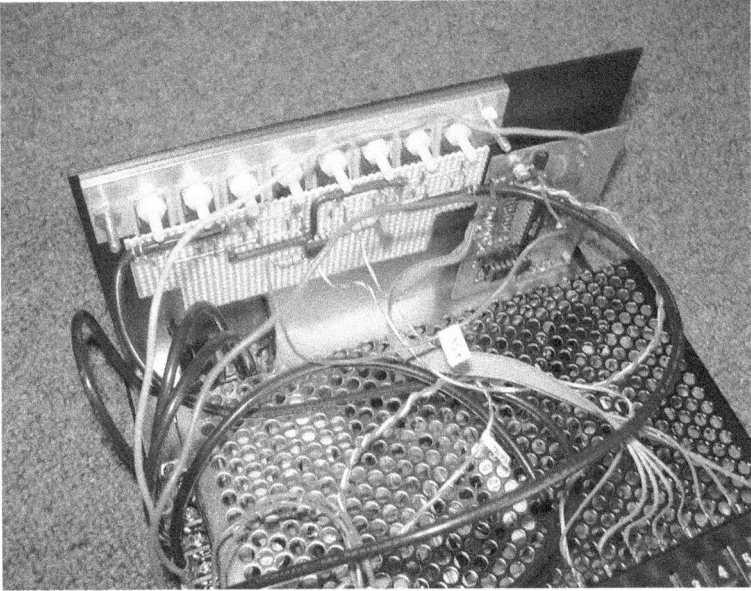

Figure 3.3: Inside view of back of pulse controller.

Figure 3.4: Inside view of front of pulse controller.

Figure 3.5: Inside view of right side of pulse controller.

Figure 3.6: Inside view of left side of pulse controller.

Figure 3.7: Front view of fully assembled pulse controller.

Figure 3.8: Back view of fully assembled pulse controller.

3.10 Pulse controller parts list

- Meanwell power supply, 12V, 150W
  Manuf#S-150-12, 4.5 inch (W) X 8 inch (L) X 2.0 inch
  (H)
  RSI Power $55.70
  or
  Peak to Peak Power, Stock # 10084 $68.10

- Enclosure and attachments

  – DB9 Extension cable, female/female, "straight through",
    pin connections: 1-5, 2-4, 3-3, 6-9, 7-8.
    DB9F-DB9F 3M, Manuf Assmann, Manuf#AK182-
    3, Digikey#AE1017-ND $6.11
    10FT,DB9(F),DB9(F), Manuf#30-9510-77, Mouser#601-
    30-9510-77, non stock $3.14

  – Enclosure, alum perf 8 inch (L) X 6 inch (W) X 4.5
    inch (H)
    Mouser#537-PERF146-BT, only black version $27.63
    Digikey L180-ND $29.64
    Manuf#PERF-146 PL, plain, direct from Manuf. $27.17

  – Power entry module, cutout hole: 26.9 x 39.1 mm
    Manuf Corcom/Tyco, Manuf#10CS1, digikey#CCM1647-
    ND $8.15
    Mouser#592-10CS1 $5.83

– Binding Post, Red, Manuf Johnson Components, Manuf#
0102-001
Digikey#J164-ND $1.80
Mouser#530-111-0102-1 $1.54

– Binding Post, Black, Manuf Johnson Components,
Manuf#111-0103-001
Digikey#J165-ND $1.80
Mouser#530-111-0103-1 $1.54

– Molex/Beau Barrier Terminal Blocks .375 PCB 8P
SINGLE, Mfr:Molex, Mfr#38720-3208
Mouser#538-38720-3208 $3.39
Digikey#WM5734-ND $5.74

– LED Panel mounts for 5mm LEDs
Any 1 of these 3:
Mouser#696-SSH-LX5090 3($0.10)
Mouser#696-SSH-LXH501 3($0.10)
Mouser#696-SSH-LX5091 3($0.10)

– LED Panel mounts for 3mm LEDs
Mouser#696-SSH-LX3050 3($0.10)

– LED red, 5mm
Digikey#67-1102-ND 3($0.10)

– 4-40 Mounting screws for power supply 4($0.05)
from any hardware store

- American Power Cord, Qualtek Electronics, Part#312008-01
  Digikey Part#Q352-ND $7.14
  or
  European Power Cord, Qualtek Electronics, Part#364002-D01
  Digikey Part#Q135-ND $6.56
  or
  Australian Power Cord, Qualtek Electronics, Part#374003-A01
  Digikey Part#374003-A01-ND $9.30

- Switcher board

  - Aluminum rectangular bar
    5.6 inch (142 mm) long x 1.0 inch (25.4 mm) wide x 1/8 inch (3.175 mm) thick
    Available in hobby supply stores where K&S metals are sold (http://www.ksmetals.com).

  - The following machine screws and nuts can be purchased from a hardware store:
    Nylon machine screws
    #6-32, 1/2 inch length 8($0.18)
    Nylon hex nuts
    #6-32 8($0.15)
    Steel machine screws #6-32, 1/2 inch length 2($0.07)
    for MOSFET bar attachment
    Steel machine screws #6-32, 3/4 inch length 2($0.07)

for Terminal strip attachment
Steel hex nuts #6-32 4($0.05)

- IRF6215 MOSFETS, Manuf International Rectifier,
  Manuf#IRF6215
  Digikey#IRF6215-ND 10($1.367)

- Thermal pads for TO-220 Mouser#532-53-77-9AC
  20($0.29)
  Aavid Thermalloy Heatsink Hardware THERMAL-
  SIL INSULATOR

- Optoisolator 4N35, 6pin dip, Manuf Fairchild, Mouser#51
  4N35 $0.30

- 330 Ohm resistor, axial leaded
  Mfr:KOA Speer, Mfr#CF1/4L331J, Mouser#660-CF1/4I
  $0.05

- 470 Ohm resistor, axial leaded
  Mfr:KOA Speer, Mfr#CF1/4L471J, Mouser#660-CF1/4I
  $0.05

- 1 uF, 50V, axial leaded ceramic cap, Z5U, 20%
  Mfr:Kemet, Mfr#C430C105M5U5CA,
  Mouser#80-C430C105M5U $0.64

- Perfboard, epoxy glass 6 inch X 17 inch
  Digikey#V1122-ND $16.67

• Microcontroller board

  - Atmel AT90S2313-10PC, onlinecomponents.com 2($5.22)
    This part is no longer manufactured by Atmel, but

is available at onlinecomponents.com. Atmel's replacement for this part is ATtiny2313, Mouser# 556-ATTINY2313V10PU. The pinout is the same. There are some differences which Atmel discusses in their application note, Replacing AT90S2313 by ATtiny2313.

– DIP socket, 20pin
 Mouser#517-4820-3004-CP $0.19

– RS232 Interface IC, DIP package, Mouser#511-ST202EBN
 2($1.30)

– DIP socket, 16pin
 Mouser#517-4816-3004-CP $0.16

– 5V, 0.1A voltage reg, 78L05, TO-92, Mfr:STMicro,
 Mfr#L78L05ABZ
 Mouser#511-L78L05ABZ $0.28

– NPN 2N3906 SMD, Mouser#MMBT3906 $0.03

– 10uF electrolytic SMD cap, Mfr:Nichicon,
 Mfr#UWX1E100MCL1GB
 Mouser#647-UWX1E100MCL1 $0.17

– 22pF radial leaded cap, Mfr:Vishay/Sprague,
 Mfr#1C10C0G220J050B
 Mouser#75-1C10C0G220J050B 2($0.21)

– 220 Ohm SMD resistor, Mfr:Xicon, Mfr#260-220
 Mouser#260-220 $0.08

– 0.1uF, 50V, SMD ceramic capacitor, Mfr#699-CL21E104MBNC
 Mouser#699-CL21E104MBNC 4($0.058)

- 1uF, 50V, SMD ceramic capacitor, Mfr:AVX,
  Mfr#12065G105ZAT2A
  Mouser#581-12065G105Z $0.66

- CONN DB9 MALE SOLDER CUP TIN, Manuf:Norcomp
  Manuf#171-009-102-021
  Digikey#2209M-ND $1.57

- 1 PCB (3" X 7.5") from Advanced Circuits (4pcb.com)
  $33.00
  This PCB combines microcontroller board and amplifier which needs to be cut in two.

Chapter 4

Sensor Coil

The purpose of the sensor coil is to acquire the proton precession signal from the sample, and to provide the amplifier with as much of the signal as possible.

4.1 Sensor coil requirements

- Easy access

  We use a solenoid geometry for the sensor coil. While other geometries such as a toroid, might be used, the solenoid allows easy access to and quick replacement of the sample.

- Low noise

  A major problem that needs to be considered in designing a sensor coil is environmental noise. The world is awash in unwanted man-made magnetic field fluctuations, and a solenoidal coil will pick up some of this. In fact the traditional AM radio antenna is just a solenoidal coil wound on a ferrite rod. The way we mitigate this problem is to use two identical sensor coils.

  The two coils are wound in opposite directions with respect to each other and connected in series. They are mounted parallel to each other so that external magnetic field noise common to both coils is cancelled out. The signal from the sample in one of the coils can then be detected with minimal interference. Both coils are large enough to accomodate a 2.0 ounce (59 ml) bottle containing the sample.

  This implies that to get a signal, you must place a sample bottle into only one of the sample coils. If you were to place a bottle in each sample coil, then a signal would be produced in each coil, and the two signals would exactly cancel each other out, resulting in no signal at all.

- No ferromagnetic or metallic materials nearby

  The magnetometer is designed to operate using the Earth's magnetic field. The coils must therefore be placed in a position where there is minimal disturbance of the field. This usually means placing the coils outdoors away from any structures containing ferromagnetic materials and away from any power lines or other sources of electromagnetic interference. The coils may be operated indoors as long as it can be verified that the magnetic field is uniform. Any non-uniformity in the field will cause the resulting signal to be degraded.

When placed outside, we recommend that the coils not be put directly on the ground, since ferromagnetic rocks and other materials in the soil may cause small localized distortions of the Earth's magnetic field. A wooden or plastic stand (no nails, screws, or fasteners) is suitable for placing the coils on.

Large metallic objects that are not ferromagnetic, such as aluminum, copper, and brass, should also be kept away from the coils. This is because they can induce eddy currents when the current in the polarization coil is turned off, corrupting the magnetic field in the region of the sample. Of course, this also means that attempting to shield the coils from magnetic noise by placing them in a metallic box will guarantee that no proton precession signal is detected.

- Wire size, Johnson noise, and Q

  The wire size used, to create the sensor coils, is an important parameter. On the one hand, you want thin wire because the more turns you have, the stronger the signal. On the other hand, you want fat wire so that the Johnson noise is low, and the quality factor (Q) is high. Johnson noise is thermal noise, and is higher for larger resistance. The Q is a measure of how ideal an inductor is, and is lower for larger resistance. We have chosen 22 AWG solid copper magnet wire. This size is a suitable fit for the constraints.

4.2   How the coils are wired together

It's important to wire the two coils together correctly. If wired incorrectly, the ambient noise will not be diminished, but amplified. The two coils are wound in opposite directions with respect to each other and connected in series as shown in figure 4.1.

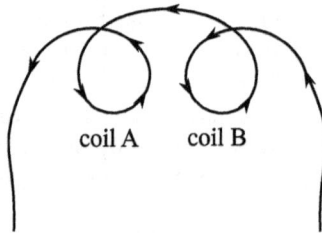

Figure 4.1: How to connect the two coils together.

Before permanently wiring them together, it's a good idea to first temporarily connect them as you think they should go, then check with an oscilloscope whether they do indeed diminish environmental noise. One way to do this is, is to place the two coils inside the polarizing coil, then connect the polarizing coil to a signal generator. If connected correctly, this signal will be greatly reduced when viewing the output of the connected coils with an oscilloscope.

4.3   How to connect cable to coils

The sample coils are connected to the amplifier via a two conductor 22 AWG shielded cable. Figure 4.2 below indicates how the cable is connected to the coils.

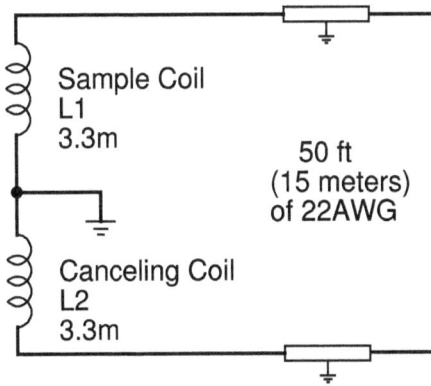

Figure 4.2: How to connect a shielded cable to the coils.

The cable shielding is connected to the ground of the amplifier, as well as the center of the two coils. The two conductors of the cable are connected to opposite ends of the coils.

4.4   Specifications of sensor coil

- 22 AWG solid copper wire, wire diameter(w/o insulation) = 0.643 mm

- coil inner diameter = 3.8 cm

- coil length = 10.0 cm

- number of wire layers = 4

- turns/layer = 138, total number of turns = 552

- Approximate inductance = 3.5 mH

- Approximate resistance (with 15 meters of line) = 5.9 Ohm

4.5   Sensor coils parts list

- 2 Acrylic tubes, 1.5 inch (3.81 cm) outer diameter, 4 inch (10.16 cm) long, 1/16 inch (0.0625 cm) thick, 1.375 inch (3.493 cm) inner diameter.

- Acrylic disk, 4.25 inch diameter, left over from polarizing coil, cut in half.

- 2 Acrylic rings, 1.7875 inch (4.54 cm) outer diameter, 1.4375 inch (3.65 cm) inner diameter, 5/32 inch (0.397 cm) thick.

- 22 AWG solid copper magnet wire, 520 feet (158.5 meter) (260 feet for each coil).

- 2 conductor shielded 22 AWG cable, 50 feet (15.2 meter).

- Spade terminals #6 for 22-18 AWG wire, Qty 3, Mouser Part#538-19131-0031.

- Plastic bottles, 2oz, translucent HDPE cylinders, at sunburstbottle.com, or specialtybottle.com.

4.6 Construction of sensor coils

In constructing the sensor coils, it is important to keep in mind that the more identical the two coils are, the better they will be at canceling environmental noise. Especially, the total number of turns should be the same. One way to check how similar the coils are, is to weigh the coil forms before they are wound, then weigh them after.

Steps for constructing the sensor coils:

1. Cut the acrylic to the dimensions given in the parts list.

2. Using an acrylic solvent, glue the coil parts together as shown in figure 3. Ensure that the sample bottle can slide into the top of at least one of the coils (the rings on the top of the tube should not get in the way of the bottle).

3. Weigh the two coil forms.

4. Wind the coils in opposite directions using 22 AWG copper magnet wire. See figure 4.1 above.

5. Weigh the two coils.

6. Compare the weight differences between the wound coil and empty coil forms. As necessary, remove wire loops to make the weight differences as identical as possible.

7. Use quick drying epoxy applied to the ends of the coil to prevent the windings from coming undone.

8. Wire the coils together so that they cancel environmental noise. See section "How the two coils are wired together".

9. Epoxy the two coils together at their base, as shown in figure 4, making sure the two coils are parallel to each other.

10. Using a scrap piece of flat acrylic, epoxy on a stabilizer (ensures that the sample coils maintain a fixed position as the polarizing coil is tilted) as shown in figure 4.5.

11. Put more epoxy around the coils as needed to make them more mechanically stable.

12. Attach the 2 conductor shielded cable as described in section 4.3.

13. Attach the spade terminals to the cable.

Figure 4.3: Sample coil forms.

Figure 4.4: Wound sample coils with sample bottle.

Figure 4.5: Sample coils with stabilizer.

Figure 4.6: Polarizing coil with sample coils inside, on tiltable platform.

Chapter 5

Amplifier

The purpose of the amplifier is to take the signal coming from the sample coil and to bring it to the level where it can be digitized by the data acquisition system. While the amplifier is less complex than the pulse controller, it is far more sensitive to how it is constructed. With a gain of 3.8 million, we are asking a lot from this amplifier, and care must be taken in its construction for it to work properly.

5.1   Why a differential amplifier is best

Figure 5.1 shows the 2 inputs to the amplifier. The ground of
the amplifier is connected to the point between the two identical
sample/canceling coils, while the opposite ends of the coils are
connected to the differential inputs of the amplifier.

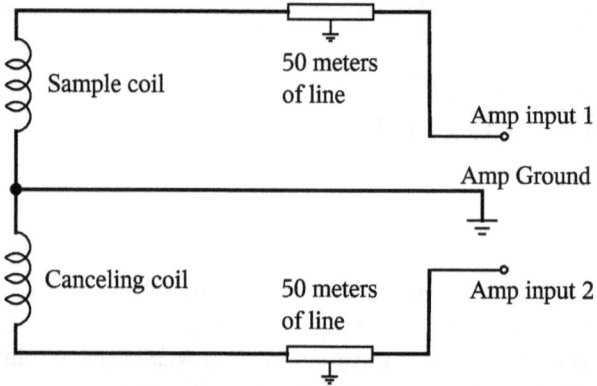

Figure 5.1: Inputs to the amplifier.

Since the sample bottle is in one of the coils, and the two coils
are otherwise identical, the two inputs will differ only in the
signal that we desire. A differential amplifier will just amplify
the difference between the two inputs, which is exactly what we
want.

5.2   Why a low noise amplifier is needed

The signal we wish to detect is in the range of a few micro-volts. We must therefore do everything possible to prevent the noise from completely swamping out the signal. This is why we use the low noise INA217 instrumentation amplifier. This instrumentation amplifier also behaves very well for low source impedance, which is what we have with the sample coils.

5.3   Description of amplifier circuit

This section describes the circuit as shown in the schematic of figure 5.2.

The first thing we see at the input of the amplifier is a bank of capacitors which are used to tune the LC circuit comprising the sample/canceling coils and those capacitors which are chosen using the dip switch. How to select these capacitor values is described in section 5.6.

The next thing after the capacitors is the 1st INA217 instrumentation amplifier. With its 5.11 Ohm resistor, it has a gain of $1 + 10K/5.11 = 1{,}958$.

The output of the 1st INA217 goes into an audio transformer. The transformer serves two purposes. The 1st purpose is to prevent DC voltage at the output of the 1st INA217 from going into

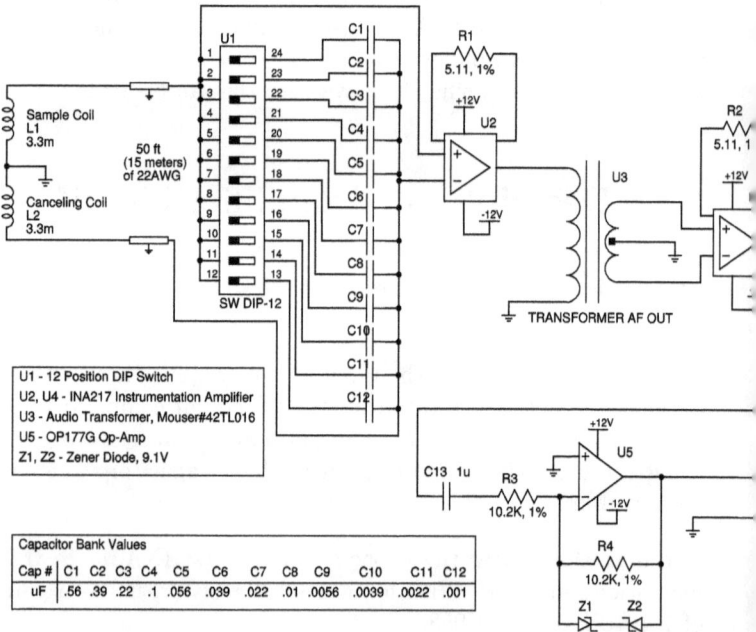

Figure 5.2: Amplifier schematic.

the input of the 2nd INA217. We don't want DC voltage going into the 2nd INA217 because that would be amplified. The 2nd purpose is to filter the signal. Our Earth's field signal will be around 2.3 kHz. The audio transformer serves as something of a bandpass filter, attenuating frequencies much greater than, and much less than, 2.3 kHz.

After the audio transformer is the 2nd INA217. The purpose of this is to simply provide more gain. With the same gain resistor as the first INA217, it also has a gain of 1,958. So the total gain after the 2nd INA217 is 1,958 ^{2} or about 3,833,000.

The ouput of the 2nd INA217 feeds into an OP177G op-amp. This op-amp circuit is set up to be a limiter-buffer. It limits the output of the amplifier to between +10 and -10 volts, protecting the input of the data acquisition system. It also serves as the workhorse providing sufficient power to drive the signal through the coax cable from the amplifier to the data acquisition system.

5.4 Supplying power to the amplifier

Another measure we must take to prevent our few microvolts of signal from getting swamped out by noise is to provide a very clean stable power source. The circuit requires +12 and -12 volts. Commercial power supplies, such as wall-warts, derived from AC sources are simply too noisy. While it may be feasible to use regulators and large capacitors to bring the noise to an acceptable level, there is an easier way.

Batteries are very clean sources of power. We use 6 volt lantern batteries. Two pair, each connected in series can provide +12 and -12 volts. The total current going into the board should be less than 50 mA, so lantern batteries ought to last for years.

Power wires going from the power supply to the circuit are like little antennas that pick up all kinds of signals from the air. It is therefore essential to use filter capacitors at the power pins of each IC. Each power pin should have a 0.1 uF or larger ceramic capacitor as near as possible to the pin, followed by a small electrolytic capacitor of 1 uF or greater value. Note that these capacitors are not shown in the schematic above.

5.5   Layout of amplifier circuit

Because of the tremendous gain of the amplifier circuit there are some **layout and construction rules** that must be followed so that you don't accidentally create an oscillator.

1. The layout of the printed circuit board must have the input on one end, and the output on the other end. That is, the signal must proceed in a straight line from input to output to minimize the possibility of feedback. If you are using the PCB file that comes with the book, then this is already done for you.

2. Power wires and signal wires going from the circuit board to connectors on the chassis must be as short as possible.

3. When the amplifier is in operation, the input cable must be as far away from the output cable as possible. They should never cross.

4. Power wires going from the battery box to the amplifier should be kept away from the input or output cables.

5. There must be a ground plane layer under the entire circuit. This is done in the PCB file that comes with the book.

6. If you have done your best to follow the rules above, and the amplifier still oscillates when you turn it on, try this. Create a wall connected to ground, using a strip of tin or copper, to physically separate the input from the output in the space inside the chassis.

5.6 Capacitors at amplifier input

A way to boost the signal, even before it gets to the instrumentation amplifiers, is to place a capacitor in series with the sample coils. Now we have an oscillator whose resonant frequency is described by

$$f = \frac{1}{2\pi\sqrt{LC}}$$

where f is the frequency in Hertz, L is the sum of the inductances of the two coils in Henrys ($\approx 2*3.5$ mH), and C is the value of the capacitor in Farads. The trick is to choose the correct capacitor value so that this oscillator has a resonant frequency tuned near the Earth's field proton precession frequency. The gain of the LC circuit will then have a maximum near the proton precession frequency.

The Earth's field can be significantly different at various locations on the Earth, and at different times of the year. In practice, we use not just one capacitor but a bank of capacitors to tune the coil. At the input of the amplifier board is a 12 position dip switch with capacitor values shown in table 5.1:

With a combination of these values, chosen with the dip switch, you should be able to get near to the Earth's field proton precession frequency at your location.

Once you get an estimate for the proton precession frequency as described in the next section, you solve for C in the formula above, then choose the combination of capacitor values in the above table which comes nearest to that value.

Table 5.1: Capacitor values for tuning sample coils.

Dip switch number	Capacitance [uF]
1	.56
2	.39
3	.22
4	.10
5	.056
6	.039
7	.022
8	.01
9	.0056
10	.0039
11	.0022
12	.001

5.7   Estimating precession frequency

Getting the resonant frequency requires us to determine the
strength of the local magnetic field. To get the strength of the
local magnetic field we need to know our latitude and longitude.
The best way to do this is with a GPS. If you do not have a
GPS then the following links may help.

- Terra Server USA (http://terraserver.microsoft.com/) -
  here you can type in an address anywhere in the United
  States to get the latitude and longitude plus maps of the
  surrounding area.

- USGS Geographic Names Information System
  (http://geonames.usgs.gov/pls/gnis/web.query.gnis.web.query
  - If your location does not have a regular street address
  you can try this website. Here you can enter for example the name of an island, a lake, a glacier or a mountain
  summit to find its latitude and longitude.

- Maporama (http://www.maporama.com/) - for areas outside the United States try this site.

Once you have your latitude and longitude, you can use a geomagnetic field model to calculate your approximate magnetic
field. There are several places on the web that will do this cal-
culation for you. NASA's National Space Science Data Center
(http://nssdc.gsfc.nasa.gov/) has an online form at

http://nssdc.gsfc.nasa.gov/space/model/models/igrf.html that you can fill in with your lat and long and it will calculate an approximate field.

An alternative to the above proceedure is to find a geomagnetic observatory near you and get some recent field measurement data from them. The U.S. Geological Survey operates several observatories. On the map at http://geomag.usgs.gov/observatories/ showing the location of these observatories, click on the one nearest you and look at their recent measurement data. A map of observatories world wide at http://www.intermagnet.org/Imomap.e.html is also available on the Intermagnet website (http://www.intermagnet.org/).

With an estimate of your local magnetic field you can calculate the proton precession frequency using the calculator at this book's web page (http://www.exstrom.com/magnum.html).

5.8 Amplifier parts list

- Instrumentation Amplifier, INA217AIDWT, 16SOIC, digikey 296-13645-1-ND 4($5.50)

- Audio Transformer Ultra-Mini 600CT-600CT, Mfr:Xicon Mfr#42TL016, mouser 42TL016 3($1.37)

- Opamp OP177G, Analog Devices, digikey OP177GSZ-ND, 2($1.75)

- BNC panel mount connector, digikey A32244-ND $2.43

- Hex nut for BNC connector, digikey A1128-ND $0.31

- Lockwasher for BNC connector, digikey A1129-ND $0.29

- 3-connector terminal block, digikey WM5740-ND $2.41

- 4 22uF SMD tantalum capacitors, digikey 399-3745-1-ND 4($0.52) = $2.08

- 4 0.1uF SMD ceramic capacitors, digikey PCC1853CT-ND 4(0.052) = $0.21

- 2 5.11 OHM resistors 1/8W 1% 0805 SMD, digikey 311-5.11CRCT-ND, 10 for $0.80

- 2 10.2K OHM resistors 1/8W 1% 0805 SMD, digikey 311-10.2KCRCT-ND, 10 for $0.80

- 1 1.0uF SMD ceramic capacitor 25V, digikey PCC2230CT-ND, 10 for $0.94

- 1 dip switch SMD 12 pos, digikey CT21912MST-ND $1.47

- 0.56uF 16V CERAMIC X7R 1206, digikey PCC1879CT-ND, 10 for $3.94

- 0.39uF 16V CERAMIC X7R 1206, digikey PCC1877CT-ND, 10 for $3.50

- 0.22uF 25V CERAMIC X7R 0805, digikey PCC1832CT-ND, 10 for $1.91

- 0.1uF 50V CERAMIC X7R 0805, digikey PCC1840CT-ND, 10 for $1.61

- 0.056uF 25V CERM X7R 0805, digikey PCC1825CT-ND, 10 for $0.62

- 0.039uF 25V CERM X7R 0805, digikey PCC1819CT-ND, 10 for $1.10

- 0.022uF 50V CERM CHIP 0805, digikey PCC223BGCT-ND, 10 for $0.37

- 0.01uF 50V CERM CHIP 0805, digikey PCC103BNCT-ND, 10 for $0.41

- 0.0056uF 50V CERM CHIP 0805, digikey PCC562BNCT-ND, 10 for $0.54

- 0.0039uF 50V CERM CHIP 0805, digikey PCC392BNCT-ND, 10 for $0.54

- 0.0022uF 50V CERM CHIP 0805, digikey PCC222BNCT-ND, 10 for $0.37

- 0.001uF 50V CERM CHIP 0805, digikey PCC102BNCT-ND, 10 for $0.43

- 3 ft of RG58 cable attached to output, digikey W300-100-ND, 3($45/100ft) = $1.35

- 4 lantern batteries Eveready, department store or batteries.com, 4($4.17) = $16.68

- 1 plastic enclosure for lantern batteries, 6 X 6 X 5 inch digikey PC-11495-ND, 9 X 8.5 X 5 inch $27.10
  Can also get cheap plastic enclosures at stacksandstacks.com

- 1 switch ROCKER DPST BLK 10A, digikey CH771-ND, cutout: 19mm X 13 mm, $1.05

- 1 enclosure, CHASSIS Alum 7 X 5 X 2 inch, digikey HM261-ND $10.11

- 1 enclosure cover, digikey HM278-ND $4.57

Figure 5.3: Amplifier with input terminal and power box.

Figure 5.4: Amplifier with output and power connector.

Figure 5.5: Amplifier inside view.

Figure 5.6: Amplifier inside view, #2.

Chapter 6

Data Acquisition and PC Control

We begin this chapter by looking at some general questions concerning data acquisition and how control and operation of the Magnum is handled by a PC. The specifics of the hardware and software used by the Magnum is then described in detail.

6.1 What sampling rate to use

The most important question regarding data acquisition is what sampling rate to use. The minimum sampling rate that you can use in acquiring any signal is twice the highest frequency component present in the signal. For example if a signal has no frequencies higher than 1000 Hz (1 kHz) then the minimum sampling rate is 2000 samples per second (2 ksps). If you sample a 1 kHz sine wave at less than 2 ksps then it will be indistinguishable from a sine wave at a lower frequency. This phenomenon is called aliasing and it can lead to spurious results if it is not prevented.

To prevent aliasing problems in your data, start by selecting a frequency, fmax, that is approximately twice the largest precession frequency that you can expect to observe. Then make sure that the input signal to the data acquisition system has no significant frequency components larger than fmax and set the sampling frequency to at least twice fmax. This will insure that the precession signal is sampled at more than twice the minimum rate and it will prevent any aliasing.

In the Earth's magnetic field for example, you should not encounter any precession frequencies outside the range of 1.5 - 2.4 kHz. You can therefore use an fmax of 5 kHz and a minimum sampling rate of 10 ksps. With the Magnum amplifier you can also be sure that no significant signals with frequencies greater than 5 kHz will make it to the data acquisition system.

6.2 What resolution to use

The next thing to consider is what resolution to use. Resolution refers to how many unique voltage levels the data acquisition system recognizes and this is determined by the number of bits used to represent each sample. What happens is that the system converts a continuous voltage signal into a fixed length binary number which can only represent a finite number of unique voltages.

Suppose for example that the data acquisition system uses 8 bits to represent each sample. It will then be able distinguish $2^8 = 256$ unique levels. This means that if the input voltage ranges from -10 to +10 volts, only voltages that differ by more than $20 / 256 = 0.078$ volts will be distinguishable.

The inability of the data acquisition system to exactly represent the input signal samples is called quantization error. This error effectively introduces noise into the data and this may decrease the overall signal to noise ratio. The required resolution, or number of bits, is therefore dependent on the characteristics of the signal and what kind of information you are trying to extract from it.

For the Magnum a resolution of 12 bits will be enough to allow the peak of the precession signal to be located with a high degree of accuracy. If you are only using the Magnum as a magnetometer then this is all the information you need. You will only need a higher resolution if you plan to use the Magnum for more general NMR type experiments which we will not discuss.

6.3   How long to take data

The length of time that data is acquired determines the degree
to which different frequencies can be resolved. The amount of
data or samples that you have effectively sets the frequency resolution. The more samples, the better the frequency resolution.

In the case of the Magnum, you should take data for a minimum
of half a second. If you take significantly less than this then
the precession signal spectrum will start to broaden and it may
become more difficult to precisely locate the peak. The usable
precession signal will last no more than 2 to 3 seconds so there
is no need to take data for much longer than this. Taking data
anywhere from 0.5 to 3.0 seconds should be sufficient for most
purposes.

6.4   The ADC board used in the Magnum

The ADC board used in the Magnum is the ADAC/5501MF
(ftp://ftp.iotech.com/ADAC.5500.20Series.20Users.20Manual.pdf)
from IOtech (http://www.iotech.com/). It is a 12 bit ADC with

8 differential (or 16 single ended) analog inputs, and a maximum sampling rate of 100 kHz.

The parts needed from IOtech are:

- Data acquisition board: ADAC/5501MF $384

- Terminal board: ADAC-TB-16 $139

- Cable: CA-G55-ADAC $69

It's not necessary to use this particular data acquisition board, but the software (for communicating with the board and getting the data) that comes with the book, is written for this board.

If you prefer to use a different data acquisition board, the essential features needed are:

1. differential inputs.

2. an external pin for triggering the acquisition of data.

3. minimum sampling rate of 10 kilosamples per second.

4. 12 bits or more per sample.

6.5 Pulse controller and data acquisition software

The overall control and operation of the Magnum is handled by a program called ADL (the name has no significance) running on a PC. The program reads a parameter file that specifies how the Magnum is to be run. The parameter file is a plain text file that can be created with any text editor.

ADL converts some of the information in the parameter file into instructions for the microcontroller in the pulse switcher. It then sends these instructions to the switcher over a serial RS232 connection. The instructions tell the switcher how long to turn on the current in the polarization coil and how long to wait between turning off the current and triggering the data acquisition.

After the switcher has been set up, ADL uses the additional information in the parameter file to set up the data acquisition system. The important data acquisition parameters are gain, sampling rate and the number of samples to take. These will all be described further below.

To start a pulse, ADL sends the run_program command to the pulse switcher. The pulse switcher then turns on the current in the polarization coil for the specified time. After turning off the current, the switcher waits a short amount of time before triggering the data acquisition system (DAQ) to start taking data.

When the data acquisition is complete, ADL will scale the data so that values reflect true voltage levels. If more than one pulse is specified then another run_program command is sent to the pulse switcher and the above process is repeated. The program averages the data from all the pulses and stores the result in a data file.

As mentioned above, the parameters that ADL needs to perform the setup and running of the system, are contained in a parameter file that is specified on the command line when the program is run. A sample parameter file is shown below.

```
ADAC5501 // Device name (32 char max)
D // Input type [D]ifferential or [S]ingle ended
0 // Channel number [0-7] (Diff) or [0-15] (SE)
B // Polarity [U]nipolar or [B]ipolar
1 // Gain [1,2,4,8,16,32,64]
16384 // Samples per second
65536 // Number of samples
6.0 // Pulse length in seconds [0-13]
0.1 // Daq trigger delay in seconds
4 // Number of pulses
10.0 // Pulse delay in seconds
magnum1.dat // Output file name
```

All these parameters must appear in every parameter file and they must appear in the order shown. The first item on each line is the value of the parameter and everything after the double slash marks is a comment. The comment describes what the parameter is and the possible values it can take.

The first seven parameters are used to initialize the data acquisition system. The next four are used for initializing the pulse programmer and running the pulse sequence. The last parameter gives the name of the data file that is created when the program is finished running. We will now describe what each of these parameters means.

The device name parameter only has meaning if there is more than one data acquisition card that is being used with the system. The parameter is actually not used by ADL but it is included so that the program can be easily modified to use more than one data acquisition card. It must however remain as the first item in the parameter file.

The input type parameter has two possible values: D for differential or S for single ended. This parameter refers to the way in which the data acquisition system treats the signals present at its input channels. With single ended inputs the voltages at all 16 input channels are digitized with respect to ground. With differential inputs what is digitized is the difference in voltage between adjacent odd and even channels. Differential inputs offer a much higher degree of noise immunity than single ended inputs. It is therefore recommended that differential inputs always be used with the Magnum.

The channel number parameter specifies the input channel of the data acquisition system that the output of the Magnum's amplifier is connected to. This is the channel that will be digitized. If you are using differential inputs (recommended) then the amplifier output should be connected to the posts marked CH(2*n) and CH(2*n+1) on the DAQ terminal board where n is the channel number that you specified. If n is 0 as in the default parameter file shown above then the two wires at the end of the coaxial cable coming from the amplifier should be connected to the posts marked CH0 and CH1.

The polarity parameter can have two values: B for bipolar and U for unipolar. Unipolar means that the input voltages to the DAQ will always be positive. Bipolar means that the input voltages can be both positive and negative. The output voltage of the Magnum amplifier will have a DC component close to zero and will swing both positive and negative so this parameter should always be set to B for bipolar.

The data acquisition system can amplify its input signal by an amount given by the gain parameter. When the input is set to bipolar it can range between $+10/g$ and $-10/g$ volts where g is the specified gain. If the input is outside this range then distortion will occur. To set this parameter first run the system with a gain of 1. The maximum value of g that you should then use must be such that your data falls within the range of $+10/g$ to $-10/g$ volts. If the Magnum is used in a magnetically noisy environment then the gain may have to be limited to 1.

The samples per second parameter specifies the number of times per second that the DAQ digitizes its input signal. As discussed above, when digitizing any signal the samples per second must be greater than twice the highest frequency component in the signal. The value of 16384 in the parameter file shown above should be more than enough for most Earth's field precession signals. Note that the value that you specify here may not be the actual value that is used by the DAQ but in most instances it should be very close. The actual sampling rate that is used will be placed in the header of the data file (see below) that is created when ADL is finished running. The actual value will also be printed out on the screen during the program's initialization of the DAQ.

The number of samples parameter simply gives the total number of times that the DAQ will digitize the signal. This is the total number of data points that will be stored in the final data file. The amount of time that this represents can be found by dividing this parameter by the samples per second parameter. For the default parameter file this is 65536/16384 = 4 seconds of data. Note however that you should use the actual sampling rate value used by the DAQ and not the value requested in the parameter file if you want an exact time.

The pulse length parameter gives the length of time in seconds that the current in the polarization coil is turned on. It therefor gives the length of time that the protons in the sample are subjected to the polarization field. The maximum time that can be specified is 13 seconds. Longer times do not provide any significant increase in performance and cause undue strain on the electronics of the switching circuit. The maximum time resolution of this parameter is 102.4 microseconds. This degree of resolution is more than enough for most uses of the Magnum. What this means is that there will be a difference between the times 6.0001 and 6.0002 but none between the times 6.0001 and 6.00011.

The DAQ trigger delay parameter specifies the amount of time between the end of the pulse and the beginning of the data acquisition. At the end of the pulse the current in the polarization coil is turned off. This induces voltages in the sample coil that will mask any precession signal. These voltages should disappear within 100 milliseconds after the current is turned off. The signal during this time period is of little use and so it makes no sense to acquire data until these turnoff transients have settled down. The 0.1 second trigger delay time specified in the default parameter file should be long enough to insure that the transients have died down. You can experiment with longer or shorter times.

Both the pulse length and DAQ trigger delay parameters are converted by ADL into instructions that can be interpreted by the pulse switcher microcontroller. These instructions are sent to the pulse switcher by ADL. The pulse switcher stores the instructions and waits for the run_program command from ADL. It is the microcontroller in the pulse switcher that actually performs the timing of the pulse and the DAQ trigger delay.

Executing multiple pulses and averaging the data obtained from each pulse can improve the signal to noise ratio of the NMR signal. The number of pulses parameter tells ADL how many pulses it should execute. Each pulse is executed using the same pulse and DAQ system parameters.

The pulse delay parameter specifies how long ADL should wait between execution of pulses. When executing multiple pulses the switching circuit in the pulse switcher may begin to heat up. This can increase the time required to turn off the current in the polarization coil which in turn may decrease the strength of the precession signal. Adding a delay between pulses allows the switching circuit to better dissipate the energy that it absorbs during the switching process. With a pulse delay, one complete pulse will last a period of time equal to the pulse length plus the DAQ trigger delay plus the data acquisition time plus the pulse delay.

The output file name is the name of the data file where ADL stores the average of the data from all the pulses that it executes. This file will have several comment lines at the beginning that are collectively called the file header. The format of a typical data file is shown below.

```
#differential :Input type
#0 :Channel number
#bipolar :Polarity
#1 :Gain
#16384.000000 :Requested samples per second
```

```
#16393.442623 :Actual samples per second
#65536 :Number of samples
#0.100000 :Daq delay
#6.000000 :Pulse length
#4 :Number of pulses
#10.000000 :Pulse delay
-4.971924
-9.237061
-8.190918
-8.251953
-7.843018
-7.907715
-1.923828
5.462646
9.324951
```

The first 11 lines constitute the file header. These lines list the parameters used by ADL to create the data. The actual number of samples per second used by the data acquisition system is also listed. The data points follow the file header. The number of data points in the above file will be 65536 since this is the number of samples requested. The values of the data points represent the actual voltage levels of the signal. In this data file 4 pulses were used so that the data values are actualy averages over the 4 pulses.

This completes the description of the software that controls the magnum. A listing of ADL is given in Appendix A.

Chapter 7

Data Processing and Analysis

7.1 Signal averaging

7.1.1 Why it improves signal to noise ratio

Signal averaging is the practice of adding together data sets taken under identical conditions. The desired effect is to reduce the noise and strengthen the signal. Signal averaging works on the principle that noise is random, while the signal embedded in the noise is not. Therefore, adding multiple data sets together tends to cancel the noise while strengthening the signal.

7.1.2   Limits due to MOSFET slow down

The pulse controller of the Magnum uses MOSFETs to turn off the current in the polarizing coils. Each time the current is turned off, the energy required to do this is dissipated in the MOSFETs, causing them to heat up. As the current is turned on and off, the MOSFETs continue to get warmer.

The turn-off time for MOSFETs increases as their temperature increases. The result is that each successive data set taken with the pulse controller has a signal that is slightly diminished in magnitude and slightly shifted later in time. The signal in each data set is therefore not identical, and signal averaging in this case is not as effective as in theory.

Note that keeping the MOSFETs thermally tied together and in contact with a good heatsink, as is done in the pulse controller, minimizes this limitation of the MOSFETs.

7.2 Filtering the data

The raw data generally contains much noise. Figure 7.1 shows a plot of the first 4 seconds of raw data, with a DAQ trigger delay of 0.1 seconds (sampling rate: 16384, sample: distilled water, Location: Longmont, Colorado, Time: Feb 8 2007, 17:34 MST).

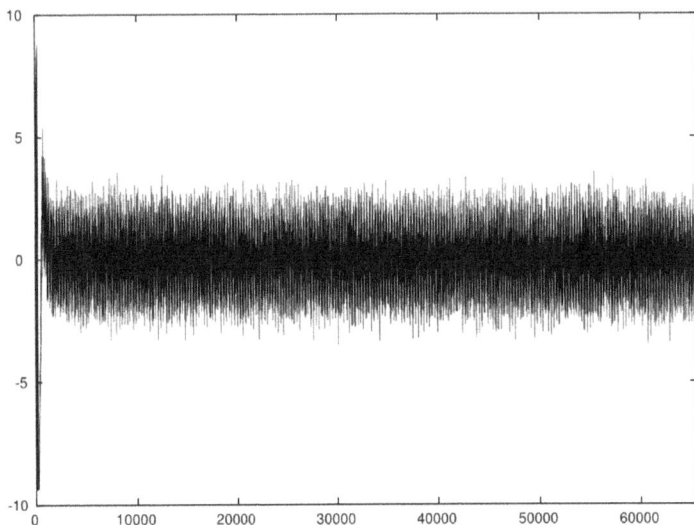

Figure 7.1: Raw data: 1st 4 seconds, trigger delay 0.1 sec, distilled water sample.

The first thing to notice is that on the left side we see transients due to the current switching. Also, it is very noisy, and the signal cannot be seen.

Let's remove the first 2000 data points to eliminate the transients, and just keep the first 2 seconds (32768 points) after that. Figure 7.2 shows the result.

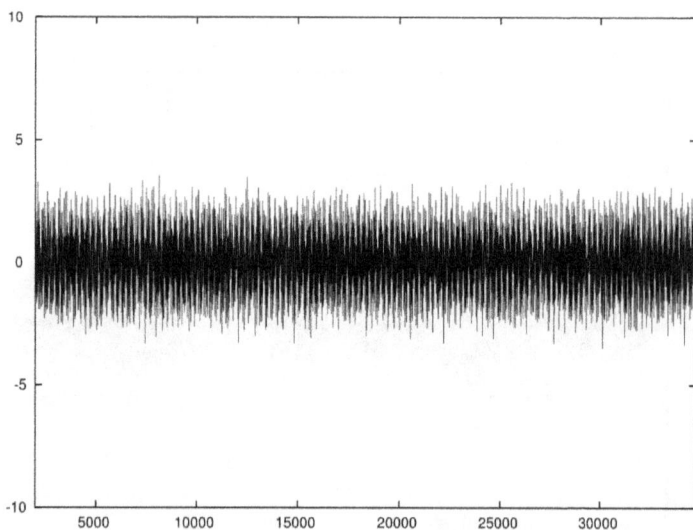

Figure 7.2: Raw data: 1st 2 seconds after removing transients.

Note: All software mentioned below can be found on our Digital Signal Processing Page (http://www.exstrom.com/journal/sigproc/)

Now let's remove some noise by applying a 6th order Butterworth bandpass filter to this data. From our estimate of the Earth's magnetic field strength as described in Chapter 5, our proton precession frequency in Longmont, Colorado should be about 2250 Hz. So we'll make the cutoff frequencies 2100 and 2400 Hz.

We create the coefficient file for our 6th order Butterworth bandpass filter like this:

```
bwbp 6 0.25620000 0.29280000 1 bpfmagnum.cof
```

where the floating point numbers are the dimensionless cutoff frequencies, converted from the actual frequencies with the calculator on our Digital Signal Processing Page. Note that to make this conversion you must know the sample rate used to take the data. If using the software that comes with this book, it's in the header of the data file. In this case it's 16393.4426 sps.

Now the frequency response of our bandpass filter is gotten with

```
rffr bpfmagnum.cof 32768 bpfmagnum.fft
```

Extracting the magnitude, which is the first column of this file (2nd column is phase):

```
extract r bpfmagnum.fft 32768 bpfmagnum.mag
```

Now we can see in figure 7.3 what the frequency response of our filter is.

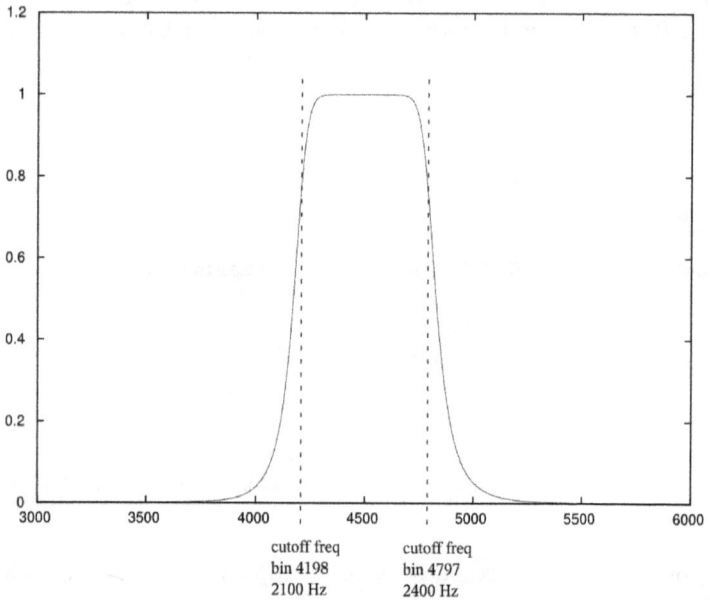

Figure 7.3: Butterworth bandbpass filter freq response, 6th order, fc = 2100, 2400 Hz.

We can now filter our Magnum data file:

`rdf bpfmagnum.cof elpaso1.dat 2000 32768 elpaso1bpf.dat`

and the result is shown in figure 7.3.

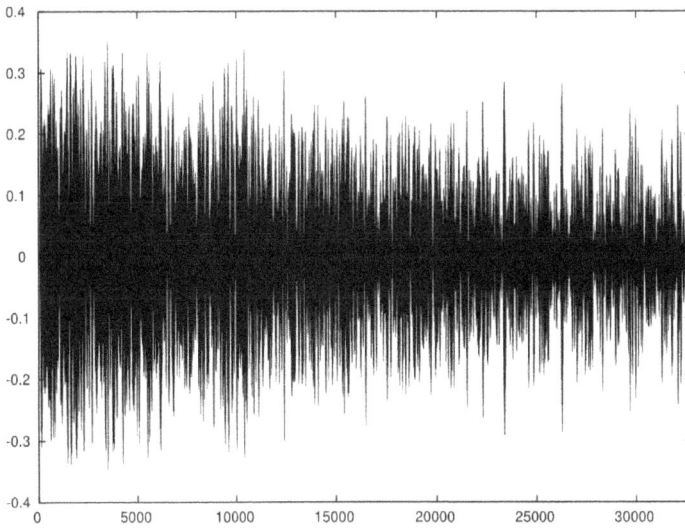

Figure 7.4: Data filtered with Butterworth BPF, 6th order.

Now we can actually see our free induction decay during two 2 seconds of data.

7.3   Spectral analysis using the FFT

Let's take the FFT of our original data and our filtered data (remember the filtered data has the first 2000 points already removed)

```
fft 2000 32768 elpaso1.dat elpaso1.fft
fft 0 32768 elpaso1bpf.dat elpaso1bpf.fft
```

Taking the magnitude of our FFT's:

```
extract m elpaso1.fft 32768 elpaso1.mag
extract m elpaso1bpf.fft 32768 elpaso1bpf.mag
```

Now we can plot and compare the FFT's, shown in figure 7.5 and 7.6

You can see how good a job our bandpass filter did.

And figure 7.7 shows a closeup view of the peak region.

You might wonder, with so much noise in the FFT of the raw data, how can you know which peak is the proton precession frequency? Simple, just remove the sample bottle, take another set of data, and the peak you're after will disappear. Of course

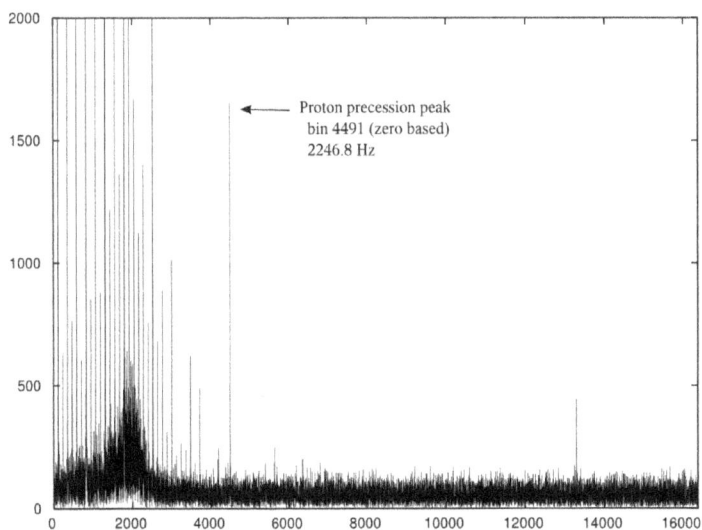

Figure 7.5: FFT of raw data, 32768 points.

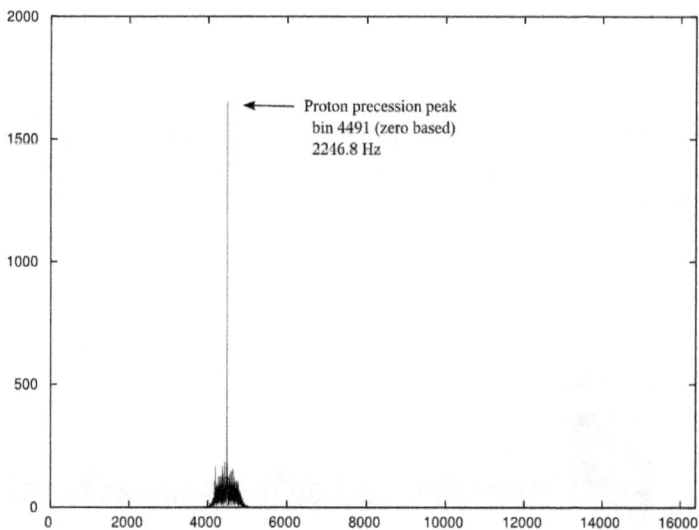

Figure 7.6: FFT of filtered data, 32768 points.

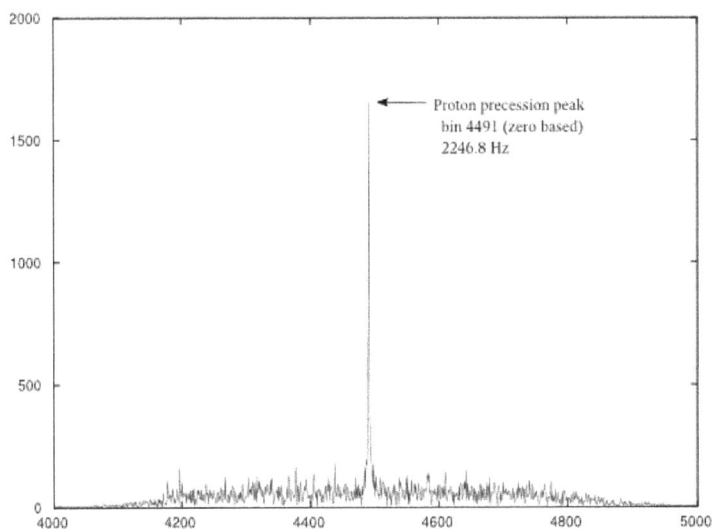

Figure 7.7: Closeup view of Figure 7.6.

you should know about where the peak is with your estimate of your local Earth's field.

If you use deionized water in place of distilled water, the peak should be significantly larger.

7.4 Converting bin number to actual frequency

Once you've identified the proton precession peak in the plot of the FFT, you can read off the bin number of the peak. The plot tells you the ballpark of the bin number, but you can actually look at the FFT magnitude data file to get the exact bin number.

In the FFT above, the peak is at data point number 4492. To get the frequency, we have to calculate the Hz/bin. From the original data file, the sampling rate is 16393.4426 sps. Our FFT used 32768 data points. We therefore have 16393.4426/32768 = 0.500288 Hz/bin. So our frequency at the peak is (4492-1)*0.500288 = 2246.7934 Hertz.

Knowing the frequency of the peak, you can now use the frequency to magnetic field converter at this book's web page (http://www.exstrom.com/magnum.html) to get the magnetic field.

7.5   High resolution spectrum for peak location

Using our hrft program
(http://www.exstrom.com/journal/sigproc/specmag.pdf) , we
can calculate a subspectrum of our data over a very small frequency range at an arbitrarily high resolution. [1] This will allow
us to determine with greater accuracy, at what frequency our
peak is.

Let's calculate 1000 Fourier transform points using our filtered
data of 32768 points:

```
hrft 2246.3 2247.3 1000 32768 16393.442623 elpaso1bpf.dat
elpaso1bpf.hrft
```

Taking the magnitude of this data set

```
extract m elpaso1bpf.hrft 1000 elpaso1bpfhrft.mag
```

the plot is shown in figure 7.8.

We find the peak to be at bin number 374 (1 based). Bin 1
is at 2246.3 Hz, bin 1000 is at 2247.3 Hz. The resolution is
therefore (2247.3 - 2246.3)/(1000 - 1) = 0.001001001 Hz/bin.
This gives our peak at frequency (374-1)*0.001001001 + 2246.3
= 2246.67337 Hz. Note that this differs from the FFT peak by

---

[1]The Goertzel program at our Digital Signal Processing Page
(http://www.exstrom.com/journal/sigproc/) can also be used to do this.

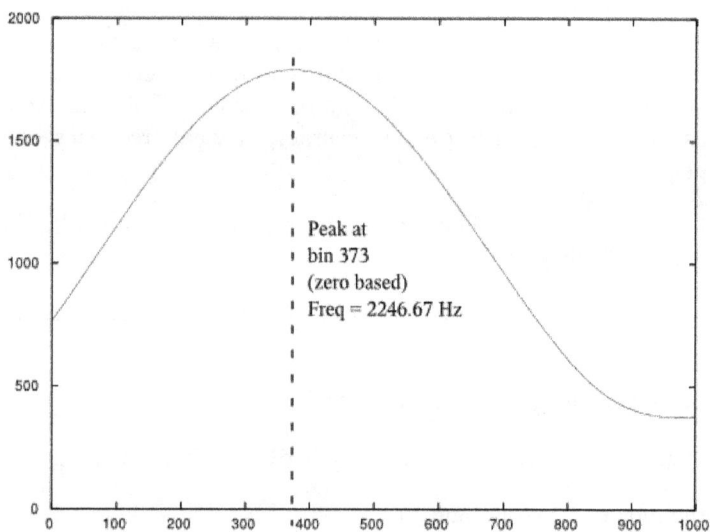

Figure 7.8: High resolution Fourier transform of filtered data, 1000 points generated, 32768 points read.

a little over 0.1 Hz.

Appendix A

Magnum Control Program

The following is a listing of program **adl.c** which is the software that handles the overall control and operation of the Magnum.

```c
#include <windows.h>
#include <conio.h>
#include <stdio.h>
#include <stdlib.h>
#include <math.h>
#include <time.h>

#define __AL_WIN32
#include "..\include\adlib.h"
#include "..\include\adlerr.h"
```

```
#define _export
#define ADLIB_CON_FILE_PATH ".\\adlib.con"

#define MAXLINESIZE 128
#define MAX_STATES  10

#define CMD_SET_PROGRAM    1
#define CMD_GET_PROGRAM    2
#define CMD_RUN_PROGRAM    3
#define CMD_LOOP_PROGRAM   4
#define CMD_STOP_PROGRAM   5

//Microcontroller clock frequency in units of Hz
#define FCLOCK_UC   10000000

HANDLE OpenSerialPort( char *portname ,DWORD baudrate );

/***************************************************************/

HANDLE OpenSerialPort( char *portname ,DWORD baudrate )
{
  HANDLE hCom;
  DWORD dwError;
  DCB dcb;
  BOOL fSuccess;

  hCom = CreateFile( portname,
    GENERIC_READ | GENERIC_WRITE,
    0,    // comm devices must be opened w/exclusive-access
    NULL, // no security attributes
    OPEN_EXISTING, // comm devices must use OPEN_EXISTING
    0,    // not overlapped I/O
    NULL  // hTemplate must be NULL for comm devices
    );
```

```
  if( hCom == INVALID_HANDLE_VALUE )
  {
    dwError = GetLastError();
    return( NULL );
  }

// Omit the call to SetupComm to use the default queue sizes.

  fSuccess = GetCommState( hCom, &dcb );
  if( !fSuccess ) return( NULL );

// Fill in the DCB: 8 data bits, no parity, 1 stop bit.

  dcb.BaudRate = baudrate;
  dcb.ByteSize = 8;
  dcb.Parity = NOPARITY;
  dcb.StopBits = ONESTOPBIT;

  fSuccess = SetCommState( hCom, &dcb );
  if( !fSuccess ) return( NULL );

  return( hCom );
}

/**************************************************************/

int main( int argc, char *argv[] )
{
  char *portname = "COMX";
  int pn;    // port number
  DWORD br; // baud rate
  HANDLE hCom;
  DWORD nbr;
  unsigned char ch;
  unsigned int nstates;
  unsigned int ntic;
```

```
FILE *fp;
char OutFile[64];
char linebuff[MAXLINESIZE];
char DevName[32];
char intype;
char polar;
int  chnum, gain, np, ix;
long i,j,ns;
double sps, asps;
double daqd,pl,pd;
double *fbuffer;

CHANLIST ChanList;
ERRNUM errnum;
LPBUFFSTRUCT lpbuff = NULL;
LHLD  lhldAdc0;  /* ADLIB device handle */

clock_t start;
double  duration;

if( argc < 3 )
{
  printf("\nUsage: adl paramfile pn\n");
  printf("  paramfile = parameter file name\n");
  printf("  pn = serial port number: 1,2,3,4\n");
  return(-1);
}


// Open the serial port connection to the micro-controler

pn = atoi( argv[2] );
if( (pn < 1) || (pn > 4) )
{
  perror( "Portnumber must be in the range [1,4]\n" );
```

```
  return( -1 );
}
portname[3] = '0' + pn;

br = 9600; //The switcher can only communicate using 9600 baud

hCom = OpenSerialPort( portname, br );
if( hCom == NULL )
{
  perror( "Could not open port\n");
  return( -1 );
}

printf( "\nSerial port opened\n" );
printf( "  Port = %s\n", portname );
printf( "  Baud = %d\n\n", br );

// Open and read the parameter file

fp = fopen(argv[1], "r");
if( fp == 0 )
{
  perror( "Unable to open parameter file" );
  return(-1);
}

printf( "Reading parameter file: %s\n\n", argv[1] );

fgets( linebuff, MAXLINESIZE, fp );
//Device name (32 char max) [ADAC5501]
sscanf( linebuff, "%32s", DevName );
fgets( linebuff, MAXLINESIZE, fp );
//Input type [D]ifferential or [S]ingle ended
sscanf( linebuff, "%c", &intype );
fgets( linebuff, MAXLINESIZE, fp );
//Channel number [0-7] (Diff) or [0-15] (SE)
```

```
sscanf( linebuff, "%d", &chnum );
fgets( linebuff, MAXLINESIZE, fp );
//Polarity [U]nipolar or [B]ipolar
sscanf( linebuff, "%c", &polar );
fgets( linebuff, MAXLINESIZE, fp );
//Gain [1,2,4,8,16,32,64]
sscanf( linebuff, "%d", &gain );
fgets( linebuff, MAXLINESIZE, fp );
//Samples per second
sscanf( linebuff, "%lf", &sps );
fgets( linebuff, MAXLINESIZE, fp );
//Number of samples
sscanf( linebuff, "%d", &ns );
fgets( linebuff, MAXLINESIZE, fp );
//Pulse length in seconds
sscanf( linebuff, "%lf", &pl );
fgets( linebuff, MAXLINESIZE, fp );
//Daq trigger delay in seconds
sscanf( linebuff, "%lf", &daqd );
fgets( linebuff, MAXLINESIZE, fp );
//Number of pulses
sscanf( linebuff, "%d", &np );
fgets( linebuff, MAXLINESIZE, fp );
//Pulse delay
sscanf( linebuff, "%lf", &pd );
fgets( linebuff, MAXLINESIZE, fp );
//Output file name (64 char max)
sscanf( linebuff, "%64s", OutFile );
fclose(fp);

printf("Read the following parameters\n\n");
printf("Device name: %s\n", DevName);
printf("Input type: %s\n",intype=='D' ? "differential":"single ended");
printf("Channel number: %d\n", chnum);
printf("Polarity: %s\n", polar == 'U' ? "unipolar" : "bipolar");
printf("Gain: %d\n", gain);
```

```
  printf("Samples per second: %f\n", sps);
  printf("Number of samples: %d\n", ns);
  printf("Pulse length (sec): %lf\n", pl);
  printf("Daq trig. delay (sec): %lf\n", daqd);
  printf("Number of pulses: %d\n", np);
  printf("Pulse delay: %lf\n", pd);
  printf("Output file name: %s\n", OutFile);

// Allocate the buffer

  fbuffer = (double *)calloc( ns, sizeof(double) );
  if( fbuffer == 0 )
  {
    printf( "Unable to allocate data buffer\n" );
    return( -1 );
  }

// Configure the pulser

  printf("\nConfiguring the pulser\n");

  if( daqd < 0 ) daqd *= -1;
  if( daqd > 6.7 ) daqd = 6.7;
  nstates = daqd > 0 ? 3 : 2;
  if( pl > 13.0 ) pl = 13.0;
  if( pl > 6.7108864 ) ++nstates;

  printf("Number of states = %1u\n", nstates);

  ch = (char)CMD_SET_PROGRAM;
  WriteFile( hCom, (LPVOID)(&ch), 1, (LPDWORD)(&nbr), NULL );

  ch = nstates & 0x000000FF;
  WriteFile( hCom, (LPVOID)(&ch), 1, (LPDWORD)(&nbr), NULL );

  ch = 0;  // first state turns current switch on
```

```
WriteFile( hCom, (LPVOID)(&ch), 1, (LPDWORD)(&nbr), NULL );

ntic = (unsigned int)(FCLOCK_UC*pl/1024.0);
if( ntic <= 65536 )
{
  ntic = 65536 - ntic;
  ch = ntic & 0x000000FF;
  WriteFile( hCom, (LPVOID)(&ch), 1, (LPDWORD)(&nbr), NULL );
  ch = (ntic & 0x0000FF00) >> 8;
  WriteFile( hCom, (LPVOID)(&ch), 1, (LPDWORD)(&nbr), NULL );
  printf("state,time = 0, %1.7lf (%1u)\n", pl, 65536-ntic);
}
else
{
  ch = 0;
  WriteFile( hCom, (LPVOID)(&ch), 1, (LPDWORD)(&nbr), NULL );
  WriteFile( hCom, (LPVOID)(&ch), 1, (LPDWORD)(&nbr), NULL );
  printf("state,time = 0, 6.7108864 (65536)\n");

  ch = 0; // One more state with current switch on
  WriteFile( hCom, (LPVOID)(&ch), 1, (LPDWORD)(&nbr), NULL );
  ntic = 2*65536 - ntic;
  ch = ntic & 0x000000FF;
  WriteFile( hCom, (LPVOID)(&ch), 1, (LPDWORD)(&nbr), NULL );
  ch = (ntic & 0x0000FF00) >> 8;
  WriteFile( hCom, (LPVOID)(&ch), 1, (LPDWORD)(&nbr), NULL );
  printf("state,time = 0, %1.7lf (%1u)\n", pl-6.7108864, 65536-ntic);
}

if( daqd > 0 )
{
  ch = 1;  // this state turns current switch off
  WriteFile( hCom, (LPVOID)(&ch), 1, (LPDWORD)(&nbr), NULL );
  ntic = (unsigned int)(FCLOCK_UC*daqd/1024.0);
  ntic = 65536 - ntic;
  ch = ntic & 0x000000FF;
```

```
    WriteFile( hCom, (LPVOID)(&ch), 1, (LPDWORD)(&nbr), NULL );
    ch = (ntic & 0x0000FF00) >> 8;
    WriteFile( hCom, (LPVOID)(&ch), 1, (LPDWORD)(&nbr), NULL );
    printf("state,time = 1, %1.7lf (%1u)\n", daqd, 65536-ntic);
}

ch = 129;   // this state triggers the data acquisition
WriteFile( hCom, (LPVOID)(&ch), 1, (LPDWORD)(&nbr), NULL );
ntic = 65534;
ch = ntic & 0x000000FF;
WriteFile( hCom, (LPVOID)(&ch), 1, (LPDWORD)(&nbr), NULL );
ch = (ntic & 0x0000FF00) >> 8;
WriteFile( hCom, (LPVOID)(&ch), 1, (LPDWORD)(&nbr), NULL );
printf("state,time = 129, 0.0001024 (1)\n");

// Configure the data acquisition system

printf("\nConfiguring the data acquisition system\n");

errnum = AL_LoadEnvironment(ADLIB_CON_FILE_PATH);
if (errnum < 0)
{
    printf("LoadEnvironment: ADLIB error#  %ld", errnum);
    return(errnum);
}

lhldAdc0 = AL_AllocateDevice("ADC0", 0);
if (lhldAdc0 < 0)
{
    printf("AllocateDevice: ADLIB error#  %ld", (ERRNUM)lhldAdc0);
    return(errnum);
}

AL_SetTriggerMode(lhldAdc0, "POST_TRIG");
AL_SetTriggerSource(lhldAdc0, "EXTERNAL");
AL_SetTrigSourceSignal(lhldAdc0, "RISING_EDGE");
```

```
AL_SetClockSource(lhldAdc0, "INTERNAL");
AL_SetClockRate(lhldAdc0, sps, AL_HERTZ);
AL_GetActualClkRate(lhldAdc0, &asps);
printf( "Requested samples per second = %lf\n", sps );
printf( "Actual samples per second = %lf\n", asps );

ChanList.lType = ARRAY_LIST;
ChanList.lNumElements = 1;
ChanList.alChannelList[0] = chnum;
ChanList.alChannelList[1] = gain;
errnum= AL_SetChannelList(lhldAdc0, &ChanList);
if (errnum < 0)
{
  printf("SetChannelList: ADLIB error#  %ld", errnum);
  return(errnum);
}

AL_SetDataOffsetGlobal(lhldAdc0,(polar=='U' ? "UNIPOLAR":"BIPOLAR"));
AL_SetBufferSize(lhldAdc0, ns);
AL_SetNumOfBuffers(lhldAdc0, 1);
AL_SetAutoInitBuffers(lhldAdc0, YES);
AL_SetBufferDoneHandler(lhldAdc0, AL_CHECK_BUFFER);

errnum = AL_InitDevice(lhldAdc0);
if (errnum < 0)
{
  printf("InitDevice: ADLIB error#  %ld", errnum);
  return(errnum);
}

printf("\nData acquisition system has been initialized\n");
printf("\nStarting pulse sequence\n\n");

for( i=0; i<np; ++i )
{
  ch = (char)CMD_RUN_PROGRAM;
```

```
WriteFile( hCom, (LPVOID)(&ch), 1, (LPDWORD)(&nbr), NULL );

printf("Executing pulse %d\n", i);

lpbuff = NULL;
AL_StartDevice(lhldAdc0);
while( lpbuff == NULL )
  lpbuff = AL_GetDoneBuffPtr(lhldAdc0);
AL_StopDevice(lhldAdc0);

printf("Saving data\n");

for( j=0; j<ns; ++j )
{
  ix = (int)(lpbuff->hpwBufferLinear[j]) & 0x0FFF;
  if( ix & 0x0800 ) ix -= 4096;
  fbuffer[j] += ix;
}

AL_ClearBufferDoneFlag(lhldAdc0, lpbuff->dwBuffNum);

printf("  Starting %lf sec. pulse delay\n", pd);
start = clock();
do
{
  duration = (double)(clock() - start) / CLOCKS_PER_SEC;
}while( duration < pd );
printf("  Pulse delay finished\n\n");
}

printf("\nPulse sequence finished\n\n");

for( j=0; j<ns; ++j )
  fbuffer[j] /= (double)np;

fp = fopen( OutFile, "w" );
```

```
if( fp != NULL )
{
  printf("Saving data to file: %s\n",OutFile);
  fprintf(fp,"# %20s:Input type\n",intype=='D'?"differential":"single en
  fprintf(fp,"# %20d:Channel number\n",chnum);
  fprintf(fp,"# %20s:Polarity\n",polar=='U'?"unipolar":"bipolar");
  fprintf(fp,"# %20d:Gain\n",gain);
  fprintf(fp,"# %20f:Requested samples per second\n",sps);
  fprintf(fp,"# %20f:Actual samples per second\n",asps);
  fprintf(fp,"# %20d:Number of samples\n",ns);
  fprintf(fp,"# %20lf:Pulse length\n",pl);
  fprintf(fp,"# %20lf:Daq triggor delay\n",daqd);
  fprintf(fp,"# %20d:Number of pulses\n",np);
  fprintf(fp,"# %20lf:Pulse delay\n",pd);

  for( i=0; i<ns; ++i )
    fprintf( fp, "%lf\n", fbuffer[i]*10.0/2048.0/(double)gain );
  fclose( fp );
}
else
  printf( "Could not open file: %s\n", OutFile );

AL_ReleaseDevice(lhldAdc0);
AL_ReleaseEnvironment();
free( fbuffer );
return( 0 );
}
```

Appendix B

Microcontroller Program

The following is a listing of the microcontroller program mag4.asm. The program is written in the assembly language of the AT90S2313. The purpose of the program is to allow the AT90S2313 to receive instructions from a PC running the ADL program and then execute those instructions. The instructions tell the AT90S2313 how long to turn on and off the current in the polarization coil and when to trigger the data acquisition system.

```
.nolist
.include "2313def.inc"
.list

.equ  MAX_STATES      = 10
```

```
.equ  CMD_SET_PROGRAM   = 1
.equ  CMD_GET_PROGRAM   = 2
.equ  CMD_RUN_PROGRAM   = 3
.equ  CMD_LOOP_PROGRAM  = 4
.equ  CMD_STOP_PROGRAM  = 5

;For the magnum port b pin 0 controls the current switch
;For the magnum port b pin 1 controls the daq board

.def  bloop=r23   ;controls looping 0 = no loop, 1 = loop.
.def  istate=r24  ;The program state counter
.def  nstate=r25  ;The number of program states

.cseg

.org $000
  rjmp  RESET
.org $005
  rjmp  TIM1_OVF     ;Timer1 Overflow interrupt handler
.org $007
  rjmp  UART_RXC     ;UART RX Complete interrupt handler

;*****************************************************
; Interrupt Handler: RESET
; Description: Execution always starts here on power up
;    or reset. There is no return from here.

RESET:

;Initialize the ports
  ser r16
  out ddrb,r16      ;make all portb pins outputs
  out ddrd,r16      ;make all portd pins outputs
                    ;  (not used for magnum)
  clr r16
  out portb,r16     ;set all portb pins low
```

```
    out portd,r16       ;set all portd pins low
                        ;  (not used for magnum)

;For the magnum bit 0 of port B controls the current switch. A low on this
;bit will turn the switch on, applying current to the coil, and a high will
;turn it off. The bit should therefor come up high so that the current is
;initially turned off.

    sbi portb,0         ;set initial current switch position to off for magnum

;Initialize stack pointer. The stack grows toward $00.
;The stack is offset from the end of the SRAM by the number of bytes needed
;to store the states. Three bytes are required to store each state.
;MAX_STATES is defined above. RAMEND is defined in: 2313def.inc

    ldi  r16,low(RAMEND-3*MAX_STATES)
    out  SPL,r16

;Initialize the general interrupt mask register to enable/disable external
;interrupts. There are two external interrupts int0 and int1 and they are
;enabled/disabled as follows:
; $00=no interrupts enabled, $40=int0, $80=int1, $C0=int0+int1
;See AT90S2313 manual p.24 for more details

    ldi r16,$00        ;no external interrupts are used in the magnum
    out GIMSK,r16

;Initialize the MCU control register.
;This can be used to control the way external interrupts are triggered,
;enable SRAM, wait states and sleep mode.
;See AT90S2313 manual p.26 for more details.
    ldi r16,$00
    out MCUCR,r16

;Initialize the timer/counter interrupt mask register.
;$80 enables just the timer1(16 bit) overflow interrupt
```

```
;See AT90S2313 manual p.24-25 for more details.
  ldi r16,$80
  out TIMSK,r16

;Initialize the timer/counter control registers.
;$00 stops the timer/counter
;See AT90S2313 manual p.31-33 for more details.
  clr r16
  out TCCR1A,r16
  out TCCR1B,r16

;Initialize the UART control register to enable transmit, receive, and
;receive interrupt. See AT90S2313 manual p.43 for more details.
  ldi   r16,$98   ;$98 = 1001 1000
  out   UCR,r16

;Initialize the UART Baud rate register using formula:
;   UBRR = fck/(16*BAUD) - 1
;fck = 10,000,000 (10 MHz), BAUD =  9600, UBRR = 64 (%error = .16)
;fck =  4,000,000 ( 4 MHz), BAUD = 19200, UBRR = 12 (%error = .16)
;See AT90S2313 manual p.44-45 for more details.
  ldi r16,64
  out UBRR,r16

;Initialize the registers
  ldi bloop,$00
  ldi istate,$00
  ldi nstate,$00

  sei  ;set the global interrupt flag

IDLE:
  nop
  rjmp IDLE

;****************************************************************
```

```
Interrupt Handler: TIM1_OVF
Description: Takes care of Timer1 overflow events.
Note: 6 clock cycles between the event and the execution of this code.

TIM1_OVF:
  clr r17
  out TCCR1B,r17   ;stop the timer

  cp istate,nstate
  brne START_NEXT_STATE
  cpi bloop,1
  breq RESET_STATES
  reti

RESET_STATES:
  clr istate
  clr r31
  ldi r30,low(RAMEND)+1

START_NEXT_STATE:
  ld r17,-Z   ;load value
  ld r18,-Z   ;load timer low byte
  ld r19,-Z   ;load timer high byte
  out portb,r17   ;set portb
  out TCNT1H,r19
  out TCNT1L,r18
  inc istate

  ldi r17,5
  out TCCR1B,r17   ;start the timer at 1024 clocks/tick
  reti

;**********************************************************************
; Interrupt Handler: UART_RXC
; Description: Takes care of UART data receive events
```

```
UART_RXC:
   in r16,UDR       ;read byte from UART data register

   clr r17
   out TCCR1B,r17   ;stop the timer

CASE0_UART:
   cpi r16,CMD_SET_PROGRAM
   brne CASE1_UART

   SET_NSTATE:
      sbis USR,RXC
      rjmp SET_NSTATE
   in nstate,UDR

   clr istate
   clr r31
   ldi r30,low(RAMEND)+1

   SET_STATE_BYTE:
      sbis USR,RXC
      rjmp SET_STATE_BYTE
   in r17,UDR

   SET_STATE_TIME0:
      sbis USR,RXC
      rjmp SET_STATE_TIME0
   in r18,UDR

   SET_STATE_TIME1:
      sbis USR,RXC
      rjmp SET_STATE_TIME1
   in r19,UDR

   st -Z,r17   ;store value
   st -Z,r18   ;store timer low byte
```

```
st -Z,r19   ;store timer high byte

inc istate
cp istate,nstate
brne SET_STATE_BYTE
reti

CASE1_UART:
  cpi r16,CMD_GET_PROGRAM
  brne CASE2_UART

  out UDR,nstate

  clr istate
  clr r31
  ldi r30,low(RAMEND)+1

GET_STATE:
  ld r17,-Z   ;load value
  ld r18,-Z   ;load timer low byte
  ld r19,-Z   ;load timer high byte

  GET_STATE_BYTE:
    sbis USR,UDRE
    rjmp GET_STATE_BYTE
  out UDR,r17

  GET_STATE_TIME0:
    sbis USR,UDRE
    rjmp GET_STATE_TIME0
  out UDR,r18

  GET_STATE_TIME1:
    sbis USR,UDRE
    rjmp GET_STATE_TIME1
  out UDR,r19
```

```
    inc istate
    cp istate,nstate
    brne GET_STATE

reti

CASE2_UART:
    cpi r16,CMD_LOOP_PROGRAM
    brne CASE3_UART

    cpi nstate,0
    brne SET_LOOP_FLAG
    reti

    SET_LOOP_FLAG:
      ldi bloop,1
      rjmp RUN_PROGRAM

CASE3_UART:
    cpi r16,CMD_RUN_PROGRAM
    brne CASE4_UART

    clr bloop

    cpi nstate,0
    brne RUN_PROGRAM
    reti

    RUN_PROGRAM:
      clr istate
      clr r31
      ldi r30,low(RAMEND)+1

      ld r17,-Z  ;load value
      ld r18,-Z  ;load timer low byte
```

```
    ld r19,-Z   ;load timer high byte

    out portb,r17   ;set portb
    out TCNT1H,r19  ;set timer high byte
    out TCNT1L,r18  ;set timer low byte
    inc istate

    ldi r17,5
    out TCCR1B,r17  ;start the timer at 1024 clocks/tick
    reti

CASE4_UART:
  cpi r16,CMD_STOP_PROGRAM
  brne CASE5_UART

  clr bloop

CASE5_UART:
  reti
```

Bibliography

[1] A. Abragam. *Principles of Nuclear Magnetism (International Series of Monographs on Physics)*. Oxford University Press, 1983.

This is one of the earliest books written on NMR but it is still one of the best. It is clear and very well written. It assumes graduate level knowledge of physics.

[2] Andreas Antoniou. *Digital filters: analysis, design, and applications*. McGraw-Hill, second edition, 1993.

This book covers all aspects of digital filters at a somewhat advanced level. Both recursive and nonrecursive filters are covered. The recursive filters include Butterworth, Chebyshev, elliptic, and Bessel types.

[3] P. T. Callaghan, C. D. Eccles, T. G. Haskell, P. J. Langhorne, and J. D. Seymour. Earth's field nmr in antarctica: a pulsed gradient spin echo nmr study of restricted diffusion in sea ice. *Journal of Magnetic Resonance*, 133(1):148–154, July 1998.

> This paper describes an Earth's field NMR spectrometer used to estimate brine content in Antarctic sea ice. The interesting thing is its similarity to high field spectrometers. It is a pulsed spectrometer, and because of the low strength of the Earth's field, it uses audio frequency pulses instead of the radio frequency pulses used in a high field spectrometer. The construction of the polarization and sensor coils is described but details on the electronics is not discussed.

[4] P. T. Callaghan and M. Le Gros. Nuclear spins in the earth's magnetic field. *American Journal of Physics*, 50(8):709–713, aug 1982.

> This paper describes an Earth's field magnetometer that is similar to the Magnum. There are crossed polarization and sensor coils as in the Magnum. It also describes how to generate spin echos by applying an audio frequency pulse across the polarization coil. This can be adapted to work with the Magnum also. The electronics is somewhat outdated but still worth looking at.

[5] Paul T. Callaghan. *Principles of nuclear magnetic resonance microscopy.* Clarendon Press, 1993.

This book is about microscopic resolution NMR imaging but it has a very good introduction to the basic principles of NMR.

[6] Wallace Hall Campbell. *Earth Magnetism : A Guided Tour Through Magnetic Fields.* Harcourt/Academic Press, 2001.

This is the most elementary introduction to geomagnetism available. It is written by a well known scientist in the field of geomagnetism. The level is basic high school with almost no mathematics. If you are completely new to geomagnetism you may want to start with this book.

[7] Wallace Hall Campbell. *Introduction to geomagnetic fields.* Cambridge University Press, second edition, 2003.

This is the best general introduction to geomagnetism. The book is at the advanced undergraduate to begining graduate student level in geophysics. The topics include mathematical modeling of the Earth's magnetic field, the effect of the sun on the field, and basic methods for measuring the field.

[8] B. P. Cowan. *Nuclear Magnetic Resonance and Relaxation.* Cambridge University Press, 1997.

This is a more modern introduction to NMR. It has a nice chapter on the analogy of NMR to the damped harmonic oscillator and its frequency response.

[9] T. J. Ericson. Nuclear magnetic resonance apparatus at low cost. *Physics Education*, 7(2):107–111, Feb 1972.

A very simple Earth's field NMR spectrometer is described. A single coil is used for both polarization an detection. The technology is early 1970s with all circuits made of discrete transistors and polarization controled by a manual switch.

[10] Frederick W. Grover. *Inductance calculations : working formulas and tables*. Instrument Society of America, 1946.

This book is packed with inductance formulas for every kind of coil imaginable. It is an invaluable reference for anyone designing coils.

[11] Paul Horowitz and Winfield Hill. *The art of electronics*. Cambridge University Press, second edition, 1989.

This is the best introduction to electronics ever written. Due to its age, some of the material related to digital and analog ICs is somewhat out of date. The material on discrete semiconductor and op-amp circuits is still the best you can find anywhere. Highly recommended.

[12] Jian-Ming Jin. *Electromagnetic Analysis and Design in Magnetic Resonance Imaging.* CRC Press, 1999.

A good presentation of the mathematics of magnetic fields and the electromagnetic analysis of coils used in MRI.

[13] Malcolm H. Levitt. *Spin dynamics: basics of nuclear magnetic resonance.* Wiley, 2001.

This book is mainly geared toward the use of NMR spectroscopy in chemistry. It has excellent introductory chapters on NMR that would be useful for anyone interested in the subject.

[14] James Mattson and Merrill Simon. *The pioneers of NMR and magnetic resonance in medicine : the story of MRI.* Bar-Ilan University Press, 1996.

This is the best historical reference on NMR. It begins with the discovery of spin and continues up to the first use of NMR imaging for medical applications by Damadian and Lauterbur. This book will give you a good overall feel for the subject of NMR. Highly recomended.

[15] Alan V. Oppenheim and Ronald W. Schafer. *Discrete-time signal processing.* Prentice Hall, 1989.

This book will give you the theoretical background for much of digital signal processing. It

includes both spectral analysis and filtering. The
level is advanced undergraduate to graduate in
engineering, physics or applied mathematics.

[16] Robert Pease. *Troubleshooting Analog Circuits.* Newnes,
1991.

This book can save you many hours of work
in troubleshooting and designing analog circuits.
The author has spent a lifetime working with
analog circuits. If you do any work in analog electronics, you should read this book.

[17] Pavel Ripka, editor. *Magnetic sensors and magnetometers.*
Artech House, 2001.

A good reference on magnetometers of all kinds.

[18] C. L. Stong. The amateur scientist. *Scientific American,*
pages 124–128, Feb 1968.

A very simple magnetometer is described that
uses the proton precession effect. The magnetometer is used to locate buried magnets in a
field. Actual magnetic field strengths cannot be
determined using this device and it is unclear how
sensitive it would be. The design is very interesting and worth looking at.

[19] G. S. Waters and G. Phillips. A new method of measuring
the earth's magnetic field. *Geophysical Prospecting,* 4:1–9,
1956.

This paper descibes what is probably one of the
first proton precession magnetometers ever built.
The basic construction is similar to the Magnum
but with 1950s technology.

[20] V. G. Welsby. *The theory and design of inductance coils.*
Wiley, second edition, 1960.

This book does not have as much information as
the Grover book. It is however more expository
and discusses some of the more subtle issues of
designing and using coils in a clear easy to understand manner.

[21] Charles S. Williams. *Designing digital filters.* Prentice-Hall,
1986.

This is the simplest introduction to digital filters
that you will find anywhere. If you have abso-
lutely no experience with digital filters or digital
signal processing, you should probably start with
this book.

[22] Se yuen Mak. Let's investigate: Six ways to measure inductance. *Physics Education*, 37(5):439, sept 2002.

This paper is an excellent introduction to measuring the inductance of coils. It requires only a
basic understanding of AC circuits and electro-
magnetism.

Index

2N3906 transistor, 38
4N35 optocoupler, 36

acrylic, 23, 25, 26, 62, 64
  solvent, 26, 29, 64
ADC, 90
ADL, 92, 93
alcohol, 12
aliasing, 88
aluminum, 59
aluminum bar, 42, 44
AM radio, 58
amplifier, 14, 62, 69, 71, 88
  circuit, 71
  differential, 70
  differential inputs, 70
  gain, 69, 71, 74
  instrumentation, 71, 79
antennas, 74
assembly language, 39

AT90S2313, 36, 38–41
ATMEL, 36
ATtiny2313, 55

batteries, 74, 81
binding post, 52
brass, 59
  rod, 26
  rods, 29
  strips, 20, 26, 29
breakdown voltage, 34, 35

capacitance
  self, 25
capacitor, 14, 75
  electrolytic, 74
capacitors, 33, 38, 80
  bank of, 71, 76
  filter, 74
channel number, 95

clock speed, 38
copper, 59
crystal, 38
current, 18, 19, 21, 22, 31, 32, 34, 36, 42
    drain, 33
    energy, 102

data acquisition, 13–15, 69, 73, 87–89, 92–95
    board, 91
differential inputs, 91, 94

Earth's magnetic field, 3, 5, 14, 20, 58, 59, 88
eddy currents, 18, 59
electric arcing, 32
electrical arc, 34
electrical isolation, 44
electromagnetic interference, 59
electrons, 9
enclosure, 42, 44, 51
epoxy, 64
Exstrom Laboratories
    Digital Signal Processing Page, 104, 113

ferromagnetic, 17, 18, 58
    rocks, 59

FFT, 108, 112, 113
    bin number, 112
    Hz/bin, 112
filter
    bandpass, 73, 105, 108
    Butterworth bandpass, 105
    cuttoff frequencies, 105
    frequency response, 105
Fourier transform, 113
    high resolution, 113
free induction decay, 108
frequency resolution, 90
frequency to magnetic field converter, 112

gain, 92
    resistor, 73
gate to source voltage, 34
gate voltage, 33, 36, 38
geomagnetic data, 20
geomagnetic field model, 78
geomagnetic observatory, 79
GMR, 4
GPS, 5
gravitational field, 9
ground plane, 75
gyromagnetic ratio, 2

Hall effect, 4

heatsink, 42, 102
hydrogen, 11

impedance
    source, 71
INA217, 71, 73
inductance, 25, 36, 62
    of a coil, 21
Intermagnet, 79
International Rectifier, 33
IRF6215's, 33, 36

lamp cord, 18
large metallic objects, 18
latitude and longitude, 78
LED, 52
limiter-buffer, 73
LM78L05 voltage regulator, 38
Longmont, Colorado, 103

magnetic compass, 5–7, 20
magnetic field, 2, 3, 7, 9, 11,
        13–15, 18, 19, 22, 59,
        112
    energy, 32
    fluctuations, 58
    local, 78, 79
    noise, 58
magnetic inclination, 20

magnetic north, 20
magnetism, 1, 12
Maporama, 78
microcontroller, 38, 41, 44
    software, 39
minimum field strength, 23
MOSFET, 33–35, 42, 102
    drain, 33, 44
    gate, 33
    on-resistance, 34, 35
    p-channel enhancement, 33
    turn-off time, 102
MRI, 4
multilayered finite length solenoid,
        22

National Space Science Data Cen-
        ter, 78
neutron, 11
noise, 5, 15, 18, 59, 102
    ambient, 60
    environmental, 63
    Johnson, 59
    low, 71
northern hemisphere, 20
nuclear magnetic resonance, 12
    NMR, 4
nuclei, 4, 11, 19
nylon, 26, 29, 44

op-amp, 73
OP177G, 73, 79
oscillator, 15, 74, 75
oscilloscope, 60

paraffin, 12
paramagnetism, 9
parameter file, 93
perfboard, 42, 54
plastic bottles, 63
plexiglass, 25
polarity, 95
polarization, 14, 15
    coil, 12, 14, 17, 18, 20, 31,
        36, 38, 59, 92
    current, 36
    field, 10, 13–15, 31
    system, 13
polarizing coil, 18, 19, 23, 25,
    29, 41, 61
power entry module, 51
power lines, 59
power supply, 36, 41, 42, 51
precess, 2, 9, 10
precession frequency, 3, 13–15,
    76, 79, 88, 105, 108
printed circuit board, 74
proton, 2, 4, 7, 9–11, 13, 14
    magnetic moments, 31

proton precession magnetome-
    ter, 2, 5
    PPM, 2, 3, 7, 12, 32
protons, 17, 19
pulse controller, 14, 15, 18, 19,
    31, 35, 38, 42, 102
pulse delay, 98
pulse length, 96
pulse program, 39, 40
pulse programmer, 94
pulse sequence, 94
Purcell, Pound, and Torrey, 12

quality factor, 60
quantization error, 89

relaxation time, 4, 10
relay, 32
resistance, 25, 34, 36, 62
resistors, 33, 80
resolution, 89
resonant frequency, 22, 75
RG58, 81
RS232, 38, 39, 44, 92

sample bottle, 58, 70
sample coil, 14, 17, 18, 69
samples per second, 96
sampling frequency, 88

sampling rate, 15, 88, 91, 92, 96, 112
    minimum, 88, 91
SCR's, 33
sensor coil, 57, 58
serial port, 38
shielded cable, 61
signal, 69, 74, 101
    Earth's field, 73
    precession, 31, 57, 88–90
signal averaging, 101
signal generator, 61
solenoid, 57
source voltage, 33, 36
spectrum, 13, 15, 90
switch, 32
    dip, 71, 76
    solid state, 33
switching speed, 42

temperature, 42
terminal block, 38, 80
Terra Server USA, 78
thermal pads, 42, 44, 54
tiltable platform, 14, 17, 26, 29
toroid, 57
transformer
    audio, 71, 73, 79
transients, 42, 104

transistors, 32
trigger delay, 97

USGS Geographic Names Information System, 78

voltage induced, 21

water, 12
    deionized, 112
    distilled, 112
wire, 18
    magnet, 60, 63, 64
    size, 59
    solid copper, 23, 62
    speaker, 24
    stranded, 42

zener diodes, 33